Leckie
the education publisher
for Scotland

T0187085

Higher
PHYSICS

Practice Workbook

ISBN 9780008446758

Published by
Leckie
An imprint of HarperCollinsPublishers
Westerhill Road, Bishopbriggs, Glasgow, G64 2QT
T: 0844 576 8126   F: 0844 576 8131
leckiescotland@harpercollins.co.uk   www.leckiescotland.co.uk

HarperCollins Publishers
Macken House, 39/40 Mayor Street Upper, Dublin 1 D01 C9W8 Ireland

This material has previously been published in the following title:
9780008263621 *Higher Physics Practice Question Book* by Paul Ferguson

Publisher: Sarah Mitchell
Project managers: Harley Griffiths,LaurenMurrayandFionaWatson

Special thanks to
QBS (layout and illustration);
Louise Robb (copy-edit); Jess White (proofread);
Peter Lindsay (answers)

Printed by Ashford Colour Press Ltd

A CIP Catalogue record for this book is available from the
British Library.

Acknowledgements
Page 13 boat image © david muscroft / Shutterstock.com
All other images © Shutterstock.com

Whilst every effort has been made to trace the copyright holders,
in cases where this has been unsuccessful, or if any have inadvertently
been overlooked, the Publishers would gladly receive any information
enabling them to rectify any error or omission at the first opportunity.

This book contains FSC™ certified paper and other controlled
sources to ensure responsible forest management.

For more information visit: www.harpercollins.co.uk/green

To access the ebook version of this Practice Workbook visit
www.collins.co.uk/ebooks
and follow the step-by-step instructions.

ANSWERS Check your answers online:
www.collins.co.uk/pages/Scottish-curriculum-free-resources

About this book

This Practice Workbook has been designed to help you feel confident about your knowledge, and about exams and assessments. It is presented in two parts to provide maximum support in both understanding and exam experience.

The topic practice section contains lots of graded practice in every single topic you will meet on your course. You can use it to consolidate your learning at any point, and to revise and refresh your knowledge in the run-up to exam time. The questions get gradually more challenging to support and extend your knowledge at the same time.

The mixed practice section then gives you the chance to put that knowledge to use in a format and standard that reflects your exams. If you get stuck on a question, you can review the relevant topic section and then come back to try it again.

Good luck!

Electron arrangements of elements

Key

Atomic number
Symbol
Electron arrangement
Name

Transition elements

Group 1 (1)	Group 2 (2)	(3)	(4)	(5)	(6)	(7)	(8)	(9)	(10)	(11)	(12)	Group 3 (13)	Group 4 (14)	Group 5 (15)	Group 6 (16)	Group 7 (17)	Group 8 (18)
1 H 1 Hydrogen																	2 He 2 Helium
3 Li 2,1 Lithium	4 Be 2,2 Beryllium											5 B 2,3 Boron	6 C 2,4 Carbon	7 N 2,5 Nitrogen	8 O 2,6 Oxygen	9 F 2,7 Fluorine	10 Ne 2,8 Neon
11 Na 2,8,1 Sodium	12 Mg 2,8,2 Magnesium											13 Al 2,8,3 Aluminium	14 Si 2,8,4 Silicon	15 P 2,8,5 Phosphorus	16 S 2,8,6 Sulfur	17 Cl 2,8,7 Chlorine	18 Ar 2,8,8 Argon
19 K 2,8,8,1 Potassium	20 Ca 2,8,8,2 Calcium	21 Sc 2,8,9,2 Scandium	22 Ti 2,8,10,2 Titanium	23 V 2,8,11,2 Vanadium	24 Cr 2,8,13,1 Chromium	25 Mn 2,8,13,2 Manganese	26 Fe 2,8,14,2 Iron	27 Co 2,8,15,2 Cobalt	28 Ni 2,8,16,2 Nickel	29 Cu 2,8,18,1 Copper	30 Zn 2,8,18,2 Zinc	31 Ga 2,18,18,3 Gallium	32 Ge 2,18,18,4 Germanium	33 As 2,18,18,5 Arsenic	34 Se 2,18,18,6 Selenium	35 Br 2,18,18,7 Bromine	36 Kr 2,18,18,8 Krypton
37 Rb 2,8,18,8,1 Rubidium	38 Sr 2,8,18,8,2 Strontium	39 Y 2,8,18,9,2 Yttrium	40 Zr 2,8,18,10,2 Zirconium	41 Nb 2,8,18,12,1 Niobium	42 Mo 2,8,18,13,1 Molybdenum	43 Tc 2,8,18,13,2 Technetium	44 Ru 2,8,18,15,1 Ruthenium	45 Rh 2,8,18,16,1 Rhodium	46 Pd 2,8,18,18,0 Palladium	47 Ag 2,8,18,18,1 Silver	48 Cd 2,8,18,18,2 Cadmium	49 In 2,8,18,18,3 Indium	50 Sn 2,8,18,18,4 Tin	51 Sb 2,8,18,18,5 Antimony	52 Te 2,8,18,18,6 Tellurium	53 I 2,8,18,18,7 Iodine	54 Xe 2,8,18,18,8 Xenon
55 Cs 2,8,18,18,8,1 Caesium	56 Ba 2,8,18,18,8,2 Barium	57 La 2,8,18,18,9,2 Lanthanoids	72 Hf 2,8,18,32,10,2 Hafnium	73 Ta 2,8,18,32,11,2 Tantalum	74 W 2,8,18,32,12,2 Tungsten	75 Re 2,8,18,32,13,2 Rhenium	76 Os 2,8,18,32,14,2 Osmium	77 Ir 2,8,18,32,15,2 Iridium	78 Pt 2,8,18,32,17,1 Platinum	79 Au 2,8,18,32,18,1 Gold	80 Hg 2,8,18,32,18,2 Mercury	81 Tl 2,8,18,32,18,3 Thallium	82 Pb 2,8,18,32,18,4 Lead	83 Bi 2,8,18,32,18,5 Bismuth	84 Po 2,8,18,32,18,6 Polonium	85 At 2,8,18,32,18,7 Astatine	86 Rn 2,8,18,32,18,8 Radon
87 Fr 2,8,18,32,18,8,1 Francium	88 Ra 2,8,18,32,18,8,2 Radium	89 Ac 2,8,18,32,18,9,2 Actinoids	104 Rf 2,8,18,32,32,10,2 Rutherfordium	105 Db 2,8,18,32,32,11,2 Dubnium	106 Sg 2,8,18,32,32,12,2 Seaborgium	107 Bh 2,8,18,32,32,13,2 Bohrium	108 Hs 2,8,18,32,32,14,2 Hassium	109 Mt 2,8,18,32,32,15,2 Meitnerium	110 Ds 2,8,18,32,32,17,1 Darmstadtium	111 Rg 2,8,18,32,32,18,1 Roentgenium	112 Cn 2,8,18,32,32,18,2 Copernicum						

Lanthanides

57 La 2,8,18,18,9,2 Lanthanum	58 Ce 2,8,18,20,8,2 Cerium	59 Pr 2,8,18,21,8,2 Praseodymium	60 Nd 2,8,18,22,8,2 Neodymium	61 Pm 2,8,18,23,8,2 Promethium	62 Sm 2,8,18,24,8,2 Samarium	63 Eu 2,8,18,25,8,2 Europium	64 Gd 2,8,18,25,9,2 Gadolinium	65 Tb 2,8,18,27,8,2 Terbium	66 Dy 2,8,18,28,8,2 Dysprosium	67 Ho 2,8,18,29,8,2 Holmium	68 Er 2,8,18,30,8,2 Erbium	69 Tm 2,8,18,31,8,2 Thulium	70 Yb 2,8,18,32,8,2 Ytterbium	71 Lu 2,8,18,32,9,2 Lutetium

Actinides

89 Ac 2,8,18,32,18,9,2 Actinium	90 Th 2,8,18,32,18,10,2 Thorium	91 Pa 2,8,18,32,20,9,2 Protactinium	92 U 2,8,18,32,21,9,2 Uranium	93 Np 2,8,18,32,22,9,2 Neptunium	94 Pu 2,8,18,32,24,8,2 Plutonium	95 Am 2,8,18,32,25,8,2 Americium	96 Cm 2,8,18,32,25,9,2 Curium	97 Bk 2,8,18,32,27,8,2 Berkelium	98 Cf 2,8,18,32,28,8,2 Californium	99 Es 2,8,18,32,29,8,2 Einsteinium	100 Fm 2,8,18,32,30,8,2 Fermium	101 Md 2,8,18,32,31,8,2 Mendelevium	102 No 2,8,18,32,32,8,2 Nobelium	103 Lr 2,8,18,32,32,9,2 Lawrencium

COMMON PHYSICAL QUANTITIES

Quantity	Symbol	Value	Quantity	Symbol	Value
Speed of light in vacuum	c	$3 \cdot 00 \times 10^8$ ms^{-1}	Planck's constant	h	$6 \cdot 63 \times 10^{-34}$ Js
			Mass of electron	m_e	$9 \cdot 11 \times 10^{-31}$ kg
Magnitude of the charge on an electron	e	$1 \cdot 60 \times 10^{-19}$C			
Universal Constant of Gravitation	G	$6 \cdot 67 \times 10^{-11}$ m^3 kg^{-1} s^{-2}	Mass of neutron	m_n	$1 \cdot 675 \times 10^{-27}$ kg
Gravitational acceleration of Earth	g	$9 \cdot 8$ ms^{-2}	Mass of proton	m_p	$1 \cdot 673 \times 10^{-27}$ kg
Hubble's constant	H_0	$2 \cdot 3 \times 10^{-18}$ s^{-1}			

REFRACTIVE INDICES

The refractive indices refer to sodium light of wavelength 589 nm and to substances at a temperature of 273 K.

Substance	Refractive index	Substance	Refractive index
Diamond	2·42	Water	1·33
Crown glass	1·50	Air	1·00

SPECTRAL LINES

Element	Wavelength /nm	Colour	Element	Wavelength /nm	Colour
Hydrogen	656	Red	Cadmium	644	Red
	486	Blue-green		509	Green
	434	Blue-violet		480	Blue
	410	Violet	Lasers		
	397	Ultraviolet	Element	Wavelength /nm	Colour
	389	Ultraviolet	Carbon dioxide	9550	Infrared
				10590	Red
Sodium	589	Yellow	Helium-neon	633	

PROPERTIES OF SELECTED MATERIALS

Substance	Density /kg m^{-3}	Melting Point /K	Boiling Point /K
Aluminium	$2 \cdot 70 \times 10^3$	933	2623
Copper	$8 \cdot 96 \times 10^3$	1357	2853
Ice	$9 \cdot 20 \times 10^2$	273	....
Sea Water	$1 \cdot 02 \times 10^3$	264	377
Water	$1 \cdot 00 \times 10^3$	273	373
Air	1·29	......	......
Hydrogen	$9 \cdot 0 \times 10^{-2}$	14	20

The gas densities refer to a temperature of 273 K and a pressure of $1 \cdot 01 \times 10^5$ pa.

Relationships required for Physics Higher

$$d = \bar{v}t$$

$$s = \bar{v}t$$

$$v = u + at$$

$$s = ut + \frac{1}{2}at^2$$

$$v^2 = u^2 + 2as$$

$$s = \frac{1}{2}(u + v)t$$

$$W = mg$$

$$F = ma$$

$$E_W = Fd$$

$$E_p = mgh$$

$$E_k = \frac{1}{2}mv^2$$

$$P = \frac{E}{t}$$

$$p = mv$$

$$Ft = mv - mu$$

$$F = \frac{Gm_1m_2}{r^2}$$

$$t' = \frac{t}{\sqrt{1 - \left(\frac{v}{c}\right)^2}}$$

$$l' = l\sqrt{1 - \left(\frac{v}{c}\right)^2}$$

$$f_o = f_s\left(\frac{v}{v \pm v_s}\right)$$

$$z = \frac{\lambda_{observed} - \lambda_{rest}}{\lambda_{rest}}$$

$$z = \frac{v}{c}$$

$$v = H_0 d$$

$$W = QV$$

$$E = mc^2$$

$$E = hf$$

$$E_k = hf - hf_0$$

$$E_2 - E_1 = hf$$

$$T = \frac{1}{f}$$

$$v = f\lambda$$

$$d\sin\theta = m\lambda$$

$$n = \frac{\sin\theta_1}{\sin\theta_2}$$

$$\frac{\sin\theta_1}{\sin\theta_2} = \frac{\lambda_1}{\lambda_2} = \frac{v_1}{v_2}$$

$$\sin\theta_c = \frac{1}{n}$$

$$I = \frac{k}{d^2}$$

$$I_1 d_1^2 = I_2 d_2^2$$

$$I = \frac{P}{A}$$

Path difference $= m\lambda$ or $\left(m + \frac{1}{2}\right)\lambda$ where $m = 0, 1, 2 \ldots$.

$$\text{random uncertainty} = \frac{\text{max.value} - \text{min.value}}{\text{number of values}}$$

$$V_{rms} = \frac{V_{peak}}{\sqrt{2}}$$

$$I_{rms} = \frac{I_{peak}}{\sqrt{2}}$$

$$Q = It$$

$$V = IR$$

$$P = IV = I^2 R = \frac{V^2}{R}$$

$$R_T = R_1 + R_2 + \ldots.$$

$$\frac{1}{R_T} = \frac{1}{R_1} + \frac{1}{R_2} + \ldots.$$

$$E = V + Ir$$

$$V_1 = \left(\frac{R_1}{R_1 + R_2}\right)V_s$$

$$\frac{V_1}{V_2} = \frac{R_1}{R_2}$$

$$C = \frac{Q}{V}$$

$$E = \frac{1}{2}QV = \frac{1}{2}CV^2 = \frac{1}{2}\frac{Q^2}{C}$$

9

Additional Relationships

Circle

circumference $= 2\pi r$

area $= \pi r^2$

Sphere

area $= 4\pi r^2$

volume $= \dfrac{4}{3}\pi r^3$

Trigonometry

$\sin \theta = \dfrac{\text{opposite}}{\text{hypotenuse}}$

$\cos \theta = \dfrac{\text{adjacent}}{\text{hypotenuse}}$

$\tan \theta = \dfrac{\text{opposite}}{\text{adjacent}}$

$\sin^2 \theta + \cos^2 \theta = 1$

the education publisher
for Scotland

Higher
PHYSICS

Topic Question Practice
Paul Ferguson

1 Motion and graphs of motion

Exercise 1A Vector addition

Example

A cross-country runner runs 2·0 km due south followed by 3·0 km south-west. Find the resultant displacement of the runner.

- First, sketch the journey using the head-to-tail rule if necessary.
- Next, use trigonometry (cosine rule here) to find the magnitude of the resultant (dotted line here).

$$a^2 = b^2 + c^2 - 2bc \, (\cos a)$$

$$a^2 = 2 \cdot 0^2 + 3 \cdot 0^2 - 2 \times 2 \cdot 0 \times 3 \cdot 0 \times (\cos 135°)$$

$$a^2 = 13 - (-8 \cdot 5)$$

$$a^2 = 21 \cdot 5$$

magnitude of resultant $(a) = 4 \cdot 6$ km

- Next, find the direction using trigonometry (sine rule here).

$$\frac{4 \cdot 6}{\sin 135°} = \frac{3}{\sin \theta}$$

$$\theta = \sin^{-1} \left(\frac{3 \sin 135°}{4 \cdot 6} \right) = 27° \text{ [provided you have included a reasonable sketch this angle will}$$

do, however if you don't include a diagram you'll have to give the direction using bearings or points of the compass: e.g. 207 or 27° west of south]

- Finally, write out the answer in full: displacement = <u>4·6 km @ 27° west of south</u>.

1 Find the resultant vector in each of the following combinations:

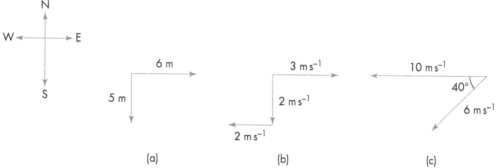

(a)   (b)   (c)

2 A sailboat travels 5·0 km due north with an average cruise speed of 4·0 m s⁻¹ before changing direction and travelling a further 3·0 km on a bearing of 030 with an average cruise speed of 3·0 m s⁻¹.

a How long does the total journey last?

b What is the resultant displacement of the sailboat?

c What is the average velocity for the whole journey?

d In which direction would the sailboat steer in order to get back to where it started?

3 An aeroplane is travelling due south with an average airspeed of 400 km h⁻¹.

The aeroplane experiences a tail wind of 50 km h⁻¹ on a bearing of 220.

Find the resultant velocity of the aeroplane as a consequence of this tail wind.

4 A canal boat is being pulled by a horse using a rope at an angle of 20° to the canal bank. The canal boat travels at a constant velocity of 1 m s⁻¹ due west, while a dog walks across the canal boat due south at 0·5 m s⁻¹.

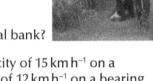

The tension in the rope is 1400 N.

a What is the component of the force in the rope parallel to the canal bank?

b Why is a long rope preferable to a short rope?

c What is the resultant velocity of the dog relative to the canal bank?

5 In a cross-country race, a runner travels with an average velocity of 15 km h⁻¹ on a bearing of 050 for 45 minutes followed by an average velocity of 12 km h⁻¹ on a bearing of 170 for 1½ hours.

a Find the total distance covered by the runner during the race.

b Find the resultant displacement of the runner.

c Calculate the average velocity of the runner.

d What bearing should be used if the runner wants to get back to the start?

6 An arrow is fired at an angle of 40° to the vertical with a speed of 44·7 m s⁻¹.

a What is the horizontal component of velocity of the arrow on take-off?

b What is the vertical component of velocity of the arrow on take-off?

7 A rope makes an angle of 20° to the horizontal and is used to drag a sled through some snow.

At one point the angle to the horizontal is increased.

What impact will this increase in angle have on:

a the horizontal component of the force being used; and

b the vertical component of the force being used?

8 During a darts match, a player throws a dart towards a dartboard. The dart lands at an angle of 70° to the vertical with a horizontal component of velocity of 5·6 m s⁻¹.

a What is the vertical component of the velocity of the dart as it hits the board?

b What is the velocity of the dart as it hits the board?

9 A golf ball is given a horizontal component of velocity of 12·0 m s⁻¹ and a vertical component of 8·60 m s⁻¹ when struck by a golf club.

a Calculate the initial speed of the ball.

b Calculate the angle the ball makes with the horizontal when it is struck.

10 A basketball enters a hoop face-on to the backboard with a vertical component of velocity of 3·6 m s⁻¹. The speed of the ball as it enters the hoop is 4·0 m s⁻¹.

a Calculate the horizontal component of velocity as the ball enters the hoop.

b Calculate the angle the ball makes with the vertical backboard as it enters the hoop.

Exercise 1B Equations of motion

Example

A car takes 4 s to accelerate constantly from rest to a speed of $12\,\mathrm{m\,s^{-1}}$.

Find the acceleration of the car.

• $u = 0\,\mathrm{m\,s^{-1}}$	first, find the appropriate relationship	$v = u + at$
• $v = 12\,\mathrm{m\,s^{-1}}$	next, insert the data into the relationship	$12 = 0 + 4a$
• $a = ?$	next, rearrange the relationship	$4a = 12 - 0$
• $t = 4\,\mathrm{s}$	finally, solve the problem	$\underline{a = 3\,\mathrm{m\,s^{-2}}}$

1 A stone is dropped from rest down a well and takes 5·5 s to reach some water.

Find the speed of the stone as it hits the water.

2 A cyclist accelerates from $2\,\mathrm{m\,s^{-1}}$ to $5\,\mathrm{m\,s^{-1}}$ in a time of 6 s.

Find the magnitude of the acceleration of the cyclist during the 6 s.

3 A car travelling at $28\,\mathrm{m\,s^{-1}}$ decelerates constantly at $5\,\mathrm{m\,s^{-2}}$ for 4 s.

Find the speed of the car after 4 s.

4 A sailboat is travelling at $8\,\mathrm{m\,s^{-1}}$ towards a finish line. A gust of wind suddenly accelerates the boat at $1·5\,\mathrm{m\,s^{-2}}$ causing its speed to increase by $6\,\mathrm{m\,s^{-1}}$ as it crosses the finish line.

For how many seconds before the boat crossed the finish line did the gust of wind last?

5 A cyclist accelerates constantly along a straight, level road at $0·75\,\mathrm{m\,s^{-2}}$ for 4 s reaching a speed of $8·0\,\mathrm{m\,s^{-1}}$.

Find the initial speed of the cyclist at the start of the acceleration.

6 A bobsleigh is approaching the finish line with a speed of $46·2\,\mathrm{m\,s^{-1}}$ and needs to stop within 65 m after crossing the finish line. Assuming the braking force is constant, find the deceleration required if the bobsleigh is to stop within the 65 m provided.

7 A car is travelling along a straight, level road at $20\,\mathrm{m\,s^{-1}}$. The brakes are applied and the car comes to rest after a further 30 m. Assuming the braking force is constant, what is the car's deceleration?

8 A bus takes 6 s to accelerate constantly from $4\,\mathrm{m\,s^{-1}}$ to $16\,\mathrm{m\,s^{-1}}$. What distance does the bus travel during the 6 s interval?

9 An astronaut is trying to repeat an experiment on the Moon, first carried out by an Apollo 15 astronaut in 1971, which proved that Galileo was correct when he claimed that all objects in the absence of air resistance will fall at the same rate towards the ground.

The astronaut drops a hammer and feather from outstretched arms and finds that they both take 1·4 s to reach the Moon's surface.

Calculate how far the hammer and feather dropped to the Moon's surface. [$g_{\mathrm{moon}} = 1·6\,\mathrm{N\,kg^{-1}}$]

10 A train is accelerating constantly at $1\cdot2\,\text{m s}^{-2}$ as it travels into a tunnel.

The speed of the train as it just enters the tunnel is $5\,\text{m s}^{-1}$.

When the rear of the train enters the tunnel it is travelling at $15\,\text{m s}^{-1}$.

Find the length of the train.

11 A train accelerates uniformly from rest out of a station to a speed of $10\,\text{m s}^{-1}$ in $10\,\text{s}$.

Find the distance the train moves in this time.

12 The largest passenger aircraft currently available is the Airbus A380. The minimum take-off speed required is $77\cdot22\,\text{m s}^{-1}$. The typical acceleration of an Airbus A380 is about $1\cdot58\,\text{m s}^{-2}$.

a What is the minimum distance required for the Airbus to lift off?

b How long will it take assuming it starts from rest?

13 A ball is thrown vertically into the air at $12\,\text{m s}^{-1}$.

a What is the maximum height reached by the ball?

b How long does the ball take to reach its maximum height?

c Find the height of the ball above the ground when it is travelling **downward** at $2\,\text{m s}^{-1}$.

14 A student drops a coin from rest down a dry well and measures a time of $4\,\text{s}$ for it to reach the bottom. On a different day the student drops another coin from rest down the well, which now has some water in it. This time the coin takes $3\,\text{s}$ to reach the top of the water.

Find the depth of the water in the well.

15 During a drag race, a car starts from rest and accelerates uniformly for $3\cdot7\,\text{s}$ along a straight level track covering a distance of $305\,\text{m}$. The car then brakes uniformly, covering a further distance of $566\,\text{m}$ before coming to rest.

a Calculate the maximum speed reached by the car.

b Calculate the time for the complete journey.

> Hint | The final speed at the end of the acceleration period equals the initial speed at the start of the deceleration period.

16 A hot air balloon is travelling upwards with a constant velocity of $2\,\text{m s}^{-1}$ when a camera is dropped over the side. The camera hits the ground $4\cdot2\,\text{s}$ later.

Find:

a the velocity of the camera *just* before striking the ground;

> Hint | The camera is still rising when it is released.

b the height of the balloon when the camera is dropped; and

c the height of the balloon when the camera reaches the ground.

17 Two students are competing against each other in an orienteering event. At the start of the course student X runs 1·2 km south (180) then 1·2 km north-west over flat ground arriving at the first check point. Student X has an average running speed of 2·8 m s⁻¹.

Student Y is better at navigating the course and runs directly from the start to the first check point with an average speed of 2·6 m s⁻¹.

a By scale drawing or otherwise, find the displacement of student X from the start point upon reaching the first check point.

b Calculate the average velocity of student X over this part of the course.

c Determine the average velocity of student Y.

d Student Y starts 6 minutes after student X. Which student arrives at the check point first? Explain your answer.

Exercise 1C Motion graphs

Example

Find the displacement of the rabbit whose motion is shown in the velocity–time graph below.

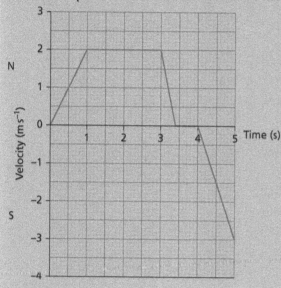

- First, write down 'displacement = area under the velocity–time graph'. This might be worth a mark in an examination.

- Next, calculate the areas above and below the time axis.

  NORTH: $(0·5 \times 2 \times 1) + (2 \times 2) + (0·5 \times 0·4 \times 2) = 5·4\,\text{m}$

  SOUTH: $(0·5 \times 1 \times 3) = 1·5\,\text{m}$

- Finally, subtract the smaller area from the larger area (direction will be in the larger of the two areas).

  $5·4 - 1·5 = \underline{3·9\,\text{m north.}}$

1 The graph of a drone flying above a field for 100 s is shown.

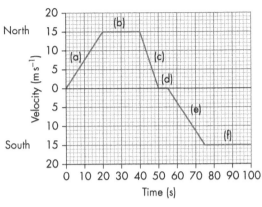

a Find the acceleration of the drone north.

b For how long was the drone hovering stationary?

c How far has the drone travelled north?

d How far has the drone travelled during the 100 s?

e What is the displacement of the drone after 100 s?

f Draw the corresponding acceleration–time graph for this drone over the first 100 s.

2 A computer-generated velocity–time graph of a bouncing ball is shown.

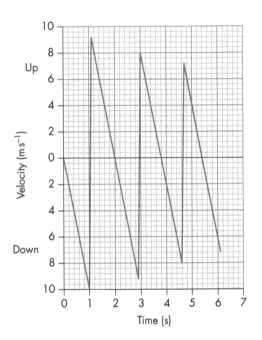

a From what approximate height is the ball dropped?

b At what time does the ball reach the ground after it is released?

c How many times does the ball bounce in the first 6 s?

d How long does the ball spend in contact with the ground during the first bounce?

e Why does the speed on take-off decrease after each bounce?

f If the mass of the ball is 50 g, how much kinetic energy is lost during the first bounce?

g At what time does the ball reach its maximum height after the second bounce?

h Why is the change in direction not simply a vertical line?

3 A cyclist is racing along a straight level track on a disused airfield.

A velocity–time graph for the cyclist is shown here.

The average velocity of the cyclist for the entire race is 7 m s^{-1}.

a Calculate the deceleration of the cyclist during part ST.

b What is the distance travelled by the cyclist up to point R?

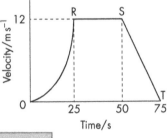

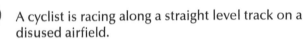
Hint You will need to calculate the total distance travelled first.

4 Another cyclist races along the same track on a disused airfield.

The velocity–time graph for this cyclist is shown here.

The distance travelled during part VW is 350 m.

Calculate the average velocity of the cyclist.

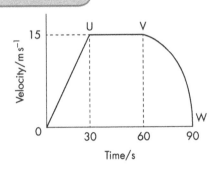

1 Motion and graphs of motion

2 Force, energy and power

Exercise 2A Unbalanced force

 1 For each of the situations below, find the unknown quantity represented by the question mark (remember vector quantities will require a direction).

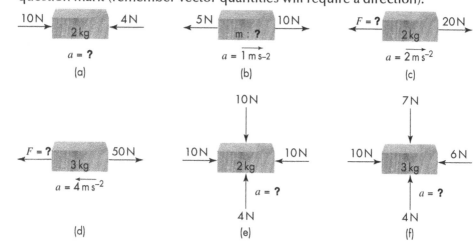

10N → 2 kg ← 4N
$a = ?$
(a)

5N ← m : ? → 10N
$a = \overrightarrow{1}\,\text{m s}^{-2}$
(b)

$F = ?$ ← 2 kg → 20N
$a = \overrightarrow{2}\,\text{m s}^{-2}$
(c)

$F = ?$ ← 3 kg → 50N
$a = \overleftarrow{4}\,\text{m s}^{-2}$
(d)

10N ↓
10N → 2 kg ← 10N
$a = ?$
↑ 4N
(e)

7N ↓
10N → 3 kg ← 6N
$a = ?$
↑ 4N
(f)

Exercise 2B Connected bodies

> ### Example
> A tractor is pulling a trailer along a straight level road by a force of 480 N as shown.
>
>
>
> F = 480 N   towbar
>
> The mass of the tractor is 1400 kg and the mass of the trailer is 1000 kg.
> Find the tension in the towbar being used to pull the trailer.

- First, find the acceleration of both tractor and trailer using Newton's second law.

$F = m_{total} \times a$

$480 = 2400 \times a$

$a = 0 \cdot 2\,\text{m s}^{-2}$

- Next, the tension in the towbar is due to the trailer only so use Newton's second law on the trailer only to find the tension in the towbar.

Tension, $T = ma = 1000 \times 0 \cdot 2 = \underline{200\,\text{N}}$

 1 Two boxes are in contact with each other on a frictionless surface. Box 'A' has a mass of 8 kg and box 'B' has a mass of 4 kg.

A 24 N unbalanced force is applied to block 'A' as shown.

a Calculate the acceleration of both blocks along the frictionless surface.

b Calculate the horizontal force acting on block 'B' alone.

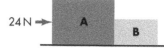

24N →

2 Two blocks of ice are in contact with each other moving horizontally along a frictionless surface with a constant acceleration of $2\,m\,s^{-2}$.

What is the force that the 3 kg block exerts on the 2 kg block?

3 A tractor has a mass of 7500 kg and is pulling a fully loaded trailer with a mass of 3500 kg along a straight level country road.

A constant frictional force of 4000 N acts on the tractor and 1500 N on the trailer.

a Calculate the forward thrust of the tractor engine when the tractor and trailer are accelerating at $1.5\,m\,s^{-2}$.

> **Hint** Find the resultant force first.

b Calculate the force exerted by the tractor on the trailer when both are accelerating at $1.5\,m\,s^{-2}$.

4 Two sleds are being pulled by a team of huskies across a frozen lake as shown below. Any friction at this point can be ignored.

The front sled (X) and passenger have a mass of 120 kg whilst the rear sled (Y) carrying supplies has a mass of 80 kg.

a What is the tension in the rope in front of the first sled (X) when the sleds are accelerating at $2\,m\,s^{-2}$?

b What is the tension in the rope connecting both sleds when accelerating at $2\,m\,s^{-2}$?

The sleds are pulled off the frozen lake onto some gravel where there is now a total frictional force of 150 N acting on both sleds.

c What is the tension in the rope in front of sled 'X' in order to maintain an acceleration of $2\,m\,s^{-2}$ across the gravel?

5 A car and caravan are travelling along a flat, straight country road with an acceleration of $4\,m\,s^{-2}$.

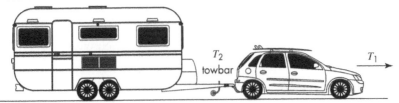

The mass of the car (including passengers) is 1300 kg and the mass of the caravan is 700 kg. Assuming initially that there are no frictional forces acting, find:

a the pulling force of the car, T_1; and

b the tension, T_2, in the towbar pulling the caravan.

c Assuming there is now an 800 N frictional force acting on both the car and the caravan, calculate the new pulling force of the car and the tension in the towbar if the car and caravan are to continue accelerating at $4\,\text{m}\,\text{s}^{-2}$.

> **Hint** Find the resultant force first.

 6 Two trolleys are joined together by a thread and pulled along horizontally by a falling 1·0 kg mass. Assume no frictional forces are acting.

Mass of trolley A = 1·5 kg

Mass of trolley B = 2·0 kg

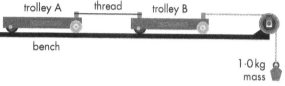
trolley A   thread   trolley B

bench

1·0 kg
mass

a Find the initial acceleration of both trolleys.

b 0·5 seconds after starting from rest the thin thread attached to both trolleys snaps.

  i Find the change of motion of trolley A after the thread snaps.

  ii Find the change of motion of trolley B after the thread snaps.

Exercise 2C Vertical motion

 1 A teacher is demonstrating Newton's laws of motion using a wooden box, a set of scales and two 5 kg masses. The apparatus is set up as shown here, with the scales attached to the top of the box.

The teacher drops the box and its contents towards the ground. What will the scales read as the box falls towards the ground? Explain.

Example

A crane is lifting two concrete blocks vertically with an acceleration of $0.2\,\text{m}\,\text{s}^{-2}$. The large block has a mass of 800 kg and the smaller block has a mass of 600 kg.

Calculate:

a the total force required to lift the two blocks; and

b the tension in the link between the blocks.

link →

- First, find the total mass being lifted and use Newton's second law to find the total force required to lift the two blocks.

Total force = unbalanced force + weight = $m_{total}a + m_{total}g = m_{total} \times (a + g)$

$$= (800 + 600) \times (0.2 + 9.8)$$

$$= \underline{14\ \text{kN}}$$

- Next, find the tension in the link, which is due only to the lower (heavier) block.

Tension = unbalanced force + weight = $ma + mg = m \times (a + g)$

$$= 800 \times (0.2 + 9.8)$$

$$= \underline{8\ \text{kN}}$$

2 A crane is used to raise an H-beam into place on a building site.

The mass of the H-beam is 2000 kg and initially it is accelerating upwards at 0·25 m s⁻².

a Calculate the tension in the cable.

b What will be the tension in the cable when the beam eventually reaches a steady speed?

cable

3 A builder uses a piece of nylon rope to raise some roof tiles vertically up onto a platform. Each roof tile has a mass of 4·4 kg and the nylon rope can withstand a force of 1264 N before it snaps.

a What is the maximum number of tiles that the builder can lift at any one time without the rope snapping?

b What is the greatest upward acceleration the builder can give the maximum number of roof tiles before the nylon rope snaps?

c What is the shortest time it can take the builder to raise the tiles through a height of 12 m?

4 A rocket with mass of 25 200 kg sits on a platform ready for take-off.

a Name a Newton pair acting in this situation before the rocket motors are fired.

When the rocket motors are fired, the rocket has an initial acceleration of 1·50 m s⁻².

b Find the thrust required from the rocket motors to accelerate the rocket at 1·50 m s⁻².

5 During a rocket flight the rocket motors exert a constant thrust of 40 kN to accelerate the rocket in outer space for 40 seconds. A graph of this acceleration is shown.

a Why does the graph take the shape shown even though the thrust from the rocket motors is constant?

b How much fuel was used during this 40-second procedure?

> **Hint** Consider the mass at the start and at the end.

c Name the Newton pair acting when the rocket is accelerating.

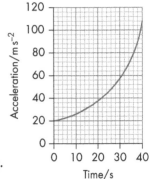

6 The motion of a lift and the tension in the support cable is recorded for three different journeys:

Motion	Tension (N)
Lift accelerating upwards	26 400
Lift moving upwards at constant speed	20 384
Lift decelerating upwards	19 440

The mass of the empty lift is 2000 kg and the average mass of a passenger is 80 kg.

a What is the initial acceleration of the lift upwards when carrying five passengers?

b Find how many passengers are in the lift when it is moving upwards with a constant speed.

c What is the initial deceleration of the lift upwards when carrying two passengers?

7  A 2 kg mass is hanging from a spring balance inside a hot air balloon that is travelling vertically upwards.
If the reading on the spring balance is 23·8 N, calculate the acceleration of the balloon.

Hint Find the unbalanced force on the spring balance.

Exercise 2D Forces at an angle

Example

A lawnmower is being pushed with a force of 100 N at an angle of 40° against a frictional force of 25·0 N. The mass of the lawnmower is 50·0 kg.

Calculate the acceleration of the lawnmower.

- First, find the force component acting horizontally $F_H = F \cos \theta$

$$= 100 \cos 40°$$
$$= 76·6 \, N$$

- Next, find the unbalanced force $= 76·6 - 25 = 51·6 \, N$

- Finally, calculate the acceleration using Newton's second law, $a = \dfrac{F}{m} = \dfrac{51·6}{50·0} = \underline{1·03 \, m\,s^{-2}}$

1 A catapult is pulled back as shown until the tension in **each** elastic cord is 34 N. Find the magnitude of the initial acceleration of a stone of mass 220 g when the elastic is released.

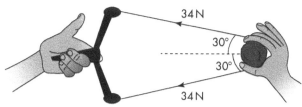

2 A water skier is being pulled along by a boat at constant speed as shown below. The tension in the rope pulling the water skier is 450 N at an angle of 25° and the skier travels parallel to the boat. Find the frictional force from the water and the air acting on the water skier.

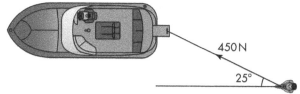

3 A volleyball net consists of a horizontal rope supported by two vertical poles.

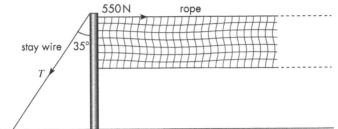

Each pole is anchored to the ground by a stay wire at an angle of 35° as shown above. The tension in the rope is 550 N. Find the tension, T, in the stay wire.

4 A pick-up truck is pulling a car of mass 1200 kg along a straight, level road by a constant force of 4000 N applied as shown at 30°. Assuming that all four wheels are in contact with the road and there is a total frictional force of 1000 N, calculate the acceleration of the car.

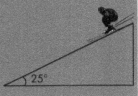

Exercise 2E Inclined planes

Example

see e.g. Exercise 2D on p. 23

A skier, mass 80·0 kg, is accelerating down a snow-covered slope at an angle of 25° to the horizontal. The frictional force acting against the skier is 40·0 N.

a Calculate the parallel component of the (weight) force acting downhill on the skier.

b Calculate the acceleration of the skier down the slope.

- First, calculate the horizontal (weight) component using $F_{\parallel} = mg \sin \theta$

$$= 80 \times 9{\cdot}8 \times \sin 25°$$
$$= 331 \, \text{N}$$

- Next, find the unbalanced force $= F_{\parallel} - F_{f} = 331 - 40{\cdot}0 = 291 \, \text{N}$

- Finally, calculate the acceleration using Newton's second law, $a = \dfrac{F}{m} = \dfrac{291}{80{\cdot}0} = \underline{3{\cdot}64 \, \text{m s}^{-2}}$

1 A cyclist freewheels with a constant velocity of 4 m s^{-1} down a slope that is inclined at 20° to the horizontal. The mass of the bicycle and rider is 90 kg. Find the force of friction acting against the bicycle and rider.

2 A sled slides down a snow-covered slope that is inclined at 15° to the horizontal. What is the initial acceleration of the sled down the slope assuming there is no friction?

3 A concrete block of mass 250 kg is pulled up a 20° incline with a constant acceleration of 1·0 m s^{-2} by a force of 2250 N. What is the frictional force opposing the motion of the block?

Hint Consider drawing a free-body diagram.

4 A 60 N force accelerates a box of mass 10 kg at 2 m s⁻² up a slope. The force of friction opposing the motion is 23 N. Find the angle of the slope.

> **Hint**   Find the unbalanced force first.

5 A child and sled (combined mass 90 kg) are being pulled up a snow-covered slope at a constant speed by the child's father. The slope is at an angle of 20° to the horizontal and a constant frictional force of 80 N acts on the sled as it moves up the slope.

a Find the parallel component of the weight force acting down the slope.

b Calculate the pulling force provided by the child's father up the slope.

When the sled reaches the top, the child and sled slide down a different slope at an angle of 15° with a constant acceleration of 2·0 m s⁻².

c Find the size of the frictional force now acting on the child and the sled.

6 A ball bearing of mass 67·6 g is projected up a slope at an angle of 25° using a loaded spring. When the ball bearing loses contact with the spring it is travelling at 6·0 m s⁻¹. The total frictional force opposing the motion of the ball bearing is 0·20 N.

a Find the deceleration of the ball bearing up the slope.

> **Hint**   Consider drawing a free-body diagram with all the forces involved after the ball bearing leaves the spring.

b Find how long it takes for the ball bearing to reach the top of the slope.

c Find the distance travelled by the ball bearing up the slope.

After reaching the top of the slope the ball bearing starts to travel back down the slope.

d Explain why the magnitude of the deceleration up the slope is different from the magnitude of the acceleration down the slope.

> **Hint**   Consider drawing a free-body diagram with all the forces involved as the ball bearing moves down the slope.

e Assuming the frictional force remains the same, find the magnitude of the acceleration down the slope.

7 A box of mass 2·0 kg sits at rest on a slope at an angle of 20° to the horizontal due to the tension applied to a string at an angle of 25° as shown. There is a force of friction acting down the slope of 1·3 N.

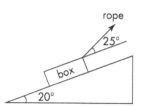

> **Hint**   Remember this is a balanced force situation, also, use the same process as that used in exercise 2D Q4.

Find the magnitude of the tension in the rope.

Exercise 2F Mechanical energy and power

Example

A warehouse worker is using a trolley to move boxes over a distance of 30 m by exerting a force of 250 N at an angle of 60° to the horizontal. The total mass of the trolley and boxes is 150 N. A constant frictional force of 40 N acts against the trolley wheels as it transports the boxes.

Find the work done by the worker.

- First, find the force component acting horizontally $F_H = F \cos \theta$

$$= 250 \cos 60°$$
$$= 125\,\text{N}$$

- Next, find the unbalanced force $\qquad\qquad = 125 - 40 = 85\,\text{N}$
- Finally, use $E_w = F \times d = 85 \times 30 = \underline{2{\cdot}55\,\text{kJ}}$

1 A pendulum bob is released at point X, which is 0·75 m above the lowest position of the bob at point Y.

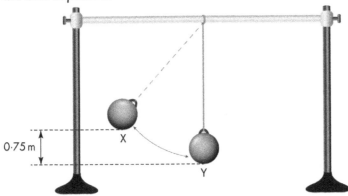

0·75 m

Assuming air resistance is negligible, what is the speed of the pendulum bob at point Y?

2 A box of mass 80 kg is raised upwards at constant speed to a height of 3 m using a fork-lift truck.

  a Calculate the work done in raising the box to the height of 3 m.

  b In practice the work done is greater than that calculated in part (a). Explain why this is so.

The box is about to be loaded onto a horizontal platform when it topples and falls towards the floor.

  c Find the speed of the box after it has fallen 2 m from rest assuming no frictional losses.

3 A small rocket moving at 240 m s⁻¹ has 720 kJ of kinetic energy. It then fires an additional rocket motor which causes it to accelerate, increasing its kinetic energy to 2·88 MJ. Calculate the new speed of the rocket.

4 A motorcycle travelling at 24·6 m s⁻¹ is brought to rest in a distance of 32 m. If the mass of the motorcycle and passenger is 500 kg, find:

  a i the kinetic energy of the motorcycle before the brakes were applied; and

    ii the work done in bringing the motorcycle to rest.

  b Calculate the average braking force required to bring the motorcycle to rest.

Example

An aluminium ball with a mass of 750g is dropped from an observation point near the top of a lighthouse. The distance fallen by the ball is 45 m.

a Find the speed of the ball as it hits the ground (assume that there is no air resistance for now).

- First, find the stored gravitational potential energy of the ball before it is dropped,

  $E_p = mgh = 0.75 \times 9.8 \times 45 = 330.75\,J$

- Next, assuming all of the stored potential energy becomes kinetic energy,

  $E_p = E_k = \frac{1}{2}\,mv^2$ i.e. $330.75 = \frac{1}{2} \times 0.75 \times v^2$

  $v^2 = 882$

  $v = \underline{29.7\,m\,s^{-1}}$

In reality, the ball experiences air resistance as it falls, which causes the ball to heat up very slightly. It also slows down slightly and now lands with a speed of $27.2\,m\,s^{-1}$.

b Find the heat energy gained by the ball as the result of the air resistance.

- First, the lost E_k between (a) and (b) will equal the heat energy gained due to friction, so calculate the new $E_k = \frac{1}{2}\,mv^2 = \frac{1}{2} \times 0.75 \times 27.2^2 = 277.44\,J$

- Finally, calculate the heat energy gained by the ball on reaching the ground,

  $E_h = 330.75 - 277.44 = \underline{53.31\,J}$

---

5 A car of mass 1200 kg is travelling at $30\,m\,s^{-1}$ along a straight level road. The brakes are applied for 3 seconds and the car decelerates to $10\,m\,s^{-1}$. How much power is lost by the car as it slows down?

6 Whilst constructing the new Queensferry Crossing bridge, a bolt is accidentally dropped. When it is falling at $4\,m\,s^{-1}$ it has 44J of kinetic energy. A short time later it has 176J of kinetic energy.

   a Find its new speed at that point.

   b Find how far it has dropped between these speeds.

7 A piece of equipment used to determine the speed of a bullet on impact is called a ballistic pendulum.

   It consists of a block of wood, mass 20·0 kg, which is suspended from a platform by two lightweight cords as shown. The bullet has a mass of 4·2g and enters the block of wood at $442\,m\,s^{-1}$ causing it to rise slightly (as shown). Find the height the block is raised by, assuming that any friction can be ignored.

8 At an airport, a suitcase of mass 22 kg slides down a slope inclined at 30° to the horizontal onto a carousel. A constant frictional force of 100 N acts against the motion of the suitcase as it travels 1·8m down the slope.

   Using only conservation of energy, calculate the speed of the suitcase when it reaches the carousel. (If you get the correct answer try the problem again using inclined planes and equations of motion to show you get the same answer.)

   Hint You need to work out the initial height to get E_p and also the work done against friction.

9 The force applied to an object over a distance of 10 m is shown on the graph.

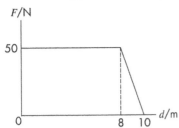

How much kinetic energy is gained over the 10 m? (Ignore any frictional forces.)

10 A trolley is released from rest down a frictionless track from a start point as shown.

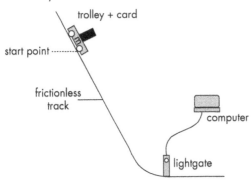

The speed of the trolley is measured at the bottom using a lightgate and a computer. How will the speed at the bottom compare if the mass of the trolley is now doubled and the trolley is released from the same height as before? Do not use any values to calculate speeds!

> **Hint** Use conservation of energy.

11 A hot-wheel toy car of mass 31 g is released from rest at point P down a frictionless track.

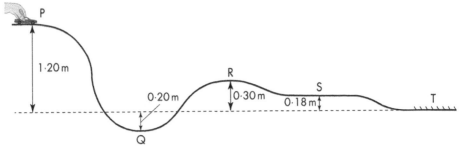

a Calculate the speed of the toy car at point R.

b At which point will the car's speed be greatest?

c Point T is carpet where the car slows down to rest. What is the frictional force acting on the car at point T if it travels a distance of 0·40 m along the carpet?

12 A car of mass 1280 kg (including two passengers) is travelling at a constant speed of 17·25 m s^{-1} along a straight level road. Ahead of the car a sheep is seen in the middle of the road and the brakes are applied bringing the car to rest. Assuming a constant braking force of 5020 N, how far will the car travel before coming to rest?

13 A motor boat is travelling across a lake with a constant speed of $2.5\,\mathrm{m\,s^{-1}}$. The frictional forces acting on the motor boat amount to $3.2\,\mathrm{kN}$.

   a Calculate the work done by the engine to counteract the frictional forces present over a distance of 100 m.

   b How much engine thrust is required to maintain the constant speed of $2.5\,\mathrm{m\,s^{-1}}$?

   c The boat engine is only 12% efficient. Calculate the total energy required by the engine over the 100 m.

14 A gondola has a mass of 1000 kg and is being used to transport 6 people up a 30° incline. At one point the gondola reaches a constant speed of $4.0\,\mathrm{m\,s^{-1}}$ and maintains this speed for a distance of 660 m. The average mass of a passenger is 75.0 kg.

   a Calculate the minimum output power of the motor which operates the gondola over this distance.

> **Hint**   Use $P = \dfrac{E_\mathrm{w}}{t}$ and remember: average speed $= \dfrac{\text{distance}}{\text{time}}$.

   b How long does it take to travel the 660 m?

   c If the gondola travels the same distance at an angle less than 30°, how will the minimum output power compare to that calculated in part (a)? You may assume the constant speed is still $4.0\,\mathrm{m\,s^{-1}}$.

   d In practice the output power required is greater than the value you calculated in part (a). Give a reason for this.

15 A technician sets up the following apparatus to demonstrate the law of conservation of energy:

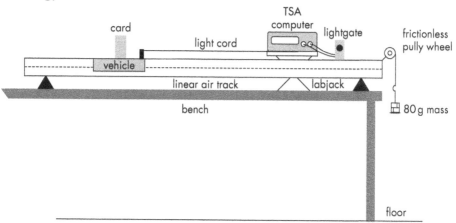

An 80 g mass is allowed to fall through 0.80 m towards the floor providing kinetic energy to a linear air track vehicle, mass 120 g, using a very light cord as shown. When the 80 g mass hits the floor the card attached to the vehicle *just* passes through a lightgate connected to a computer that is used to record the vehicle's speed.

   a Calculate the speed of the vehicle as the mass hits the floor.

> **Hint**   Ask yourself, what has acquired kinetic energy?

   b Why is it essential that the card *just* clears the lightgate when the mass hits the floor?

   c How does this set-up help verify the conservation of energy?

   d How could the procedure be improved?

16 A skier is being towed up an incline at constant speed by a tow cable attached to a pulley motor.

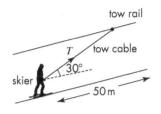

The force, T, exerted by the tow cable on the skier is 500 N and friction can be ignored. Calculate the work done by the pulley motor in pulling the skier 50 m up the slope.

17 A toy car is released from the top of a curved piece of track (point A) and it follows a loop-the-loop path as shown.

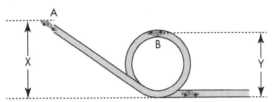

Using only the symbols m, g, x and y determine an expression for the kinetic energy of the car at point B.

18 A lightweight nylon wire is suspended between two supports. A 300 g mass is then placed in the centre of the wire causing it to sag as shown.

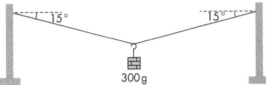

Calculate the tension in the nylon wire on both sides of the hanging mass.

> **Hint** The tension upwards from both sides of the wire counteracts the weight downwards – use trigonometry to find this tension.

19 An electric motor attached to a pulley is used to raise a concrete block which has a mass of 10·0 kg through a height of 7·5 m.

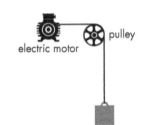

a Calculate the useful energy supplied by the motor.

b In reality the electrical energy supplied by the motor is 860 J. Find the energy wasted in lifting the block.

c Calculate the efficiency of the electric motor.

3 Momentum and impulse

Exercise 3A Momentum

> **Assume friction is negligible throughout.**

1 State the law of conservation of linear momentum.

2 A car of mass 1200 kg is travelling at 15 m s^{-1} due west. Calculate the momentum of the car.

3 A motorcycle is travelling at a constant velocity of 10 m s^{-1} due west and has a momentum of 3600 kg m s^{-1} due west. Calculate the mass of the motorcycle.

4 A bus of mass 12 600 kg has a momentum of 63 000 kg m s^{-1} due north. What is the velocity of the bus at this instant?

Example

Two linear air track vehicles are moving towards each other. Vehicle 1 has a mass of 250 g and is travelling at 0·50 m s^{-1} to the right. Vehicle 2 has a mass of 400 g and is travelling at 0·40 m s^{-1} to the left.

linear air track

Upon collision the two vehicles stick together.

a Calculate the common speed of both vehicles immediately after the collision.

- First, write out the appropriate equation: Total momentum before = Total momentum after

$$m_1u_1 + m_2u_2 = (m_1 + m_2)v$$

- Next, insert the data into the correct place $(0.25 \times 0.5) + (0.40 \times (-0.4)) = (0.25 + 0.40)v$ and remember that a direction must be assigned. **Here L→R positive.**

- Next, do the arithmetic and solve for 'v'

$$0.125 - 0.16 = 0.65v$$
$$-0.035 = 0.65v$$
$$v = -0.054 \,\text{m s}^{-1}$$

- Finally, write out the answer and don't forget the direction: $v = 0.054 \,\text{m s}^{-1}$ to the left.

b Is the collision elastic or inelastic? Explain.

- First, calculate the total kinetic energy before and after (for scalars direction is not important):

$$E_k \text{(before)} = \frac{1}{2}m_1u_1^2 + \frac{1}{2}m_2u_2^2 = \left(\frac{1}{2} \times 0.25 \times 0.5^2\right) + \left(\frac{1}{2} \times 0.40 \times 0.4^2\right) = 0.06325 \,\text{J}$$

$$E_k \text{(after)} = \frac{1}{2}(m_1 + m_2)v^2 = \frac{1}{2}(0.25 + 0.40)\,0.054^2 = 0.0009477 \,\text{J}$$

- Finally, compare the E_k before and after. Since E_k before > E_k after, collision is <u>inelastic</u>.

5 A trolley of mass 1·5 kg is travelling with a speed of 1·0 m s^{-1}. The trolley collides and sticks to a stationary trolley of mass 2·0 kg.

a Calculate the velocity of the trolleys immediately after the collision.

b Show that the collision is inelastic.

6 A 7·5 g bullet travelling at 380 m s⁻¹ collides and embeds itself in an 8·0 kg block of wood that is at rest but free to slide across a smooth surface. Find the magnitude of the velocity of the block immediately after impact.

7 Two linear air track vehicles, X and Y, are travelling towards each other as shown here.

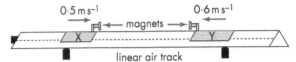

Vehicle X has a mass of 0·6 kg and vehicle Y has a mass of 0·4 kg. Upon collision, vehicle Y rebounds to the right with a velocity of 0·2 m s⁻¹. What happens to vehicle X?

8 A large bowling ball which has a mass of 6·0 kg moves at 2·0 m s⁻¹ down a lane towards a single stationary bowling pin of mass 1·5 kg. When they collide head-on, the pin moves off with a speed of 4·5 m s⁻¹. Find the speed of the bowling ball immediately after the collision.

9 A soldier on manoeuvres fires a bazooka shell, mass 1·6 kg, from a bazooka, mass 5·8 kg.

The shell travels from right to left with a muzzle velocity of 76·2 m s⁻¹. Calculate the recoil velocity of the bazooka.

10 A trolley of mass 1·5 kg is moving at a constant speed when it collides and sticks to a second stationary trolley. The graph below shows how the speed of the 1·5 kg trolley varies with time.

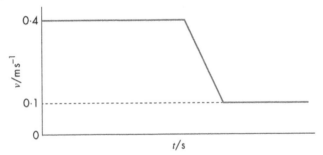

Determine the mass of the second trolley.

11 A red car, mass 1000 kg, which is travelling due east along an icy road, crashes into a stationary blue car, mass 1200 kg. Both cars lock together on impact and move off due east. Expert examination by the police establishes that both cars moved off at 8·2 m s⁻¹ immediately after the collision. Did the red car exceed the speed limit of 13·4 m s⁻¹ immediately before the collision? Justify your answer.

12 A cannon of mass 1200 kg fires a cannonball of mass 8·5 kg. The velocity with which the cannonball leaves the cannon is 65 m s⁻¹ to the right. Calculate the velocity of the cannon immediately after firing.

> **Hint** Remember that momentum is a vector quantity and direction is **very** important!

13 An ice skater of mass 75 kg moving to the right collides head-on with another skater of mass 65 kg moving at 2 m s⁻¹ to the left. The two skaters hold on to one another after the collision and move off to the left at 0·42 m s⁻¹. What was the velocity of the 75 kg skater immediately before the collision?

14 Two linear air track vehicles are travelling in the same direction as shown here.

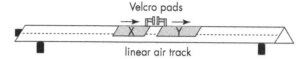

Velcro pads

linear air track

Vehicle 'X' has a mass of 1·0 kg and is travelling at 0·4 m s⁻¹. Vehicle 'Y' has a mass of 2·0 kg and is travelling at 0·1 m s⁻¹. When the two vehicles collide the Velcro pads stick to each other. How much kinetic energy is lost during the collision?

15 In a game of lawn bowls, a bowl of mass 1·35 kg is travelling with a speed of 1·20 m s⁻¹ when it hits a stationary jack of mass 250 g 'head-on'. After the collision the bowl continues to move straight on with a speed of 0·90 m s⁻¹.

a What is the speed of the jack immediately after the collision?

b How much kinetic energy is lost during the collision?

16 A missile of mass 15 kg is travelling horizontally at 280 m s⁻¹ when it suddenly explodes into two parts. One part has a mass of 6·5 kg and continues in the original direction with a speed of 120 m s⁻¹ immediately after separation. The other part also continues in the same direction. Calculate the speed of the second part immediately after the separation.

17 A 250 g vehicle travelling at 2·5 m s⁻¹ on a linear air track collides with a 480 g stationary vehicle.

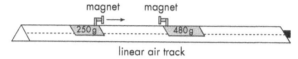

magnet   magnet

250 g   480 g

linear air track

Each vehicle is fitted with magnets that repel each other. Immediately after the interaction the 250 g vehicle repels at 1·5 m s⁻¹. What is the speed of the 480 g vehicle immediately after the interaction?

18 A curler, mass 62 kg, is sliding along the ice holding a curling stone, mass 18 kg, at 2·2 m s⁻¹ when the stone is released. The stone takes off with a speed of 3·4 m s⁻¹. Calculate the speed of the curler immediately after the stone is released.

19 Two ice skaters are practising manoeuvres on ice. The man has a mass of 75 kg and the woman has a mass of 55 kg. At one point they stand stationary next to each other and the woman pushes the man away from herself. As a result of the push the man moves off with a speed of 0·5 m s⁻¹.

What is the velocity of the woman as a result of the push?

20 A shell of mass 12 kg is travelling horizontally at a constant speed of 250 m s⁻¹ when it suddenly explodes into two parts. One part has a mass of 7 kg and continues in the original direction with a speed of 295 m s⁻¹. The other part also continues in the original direction. Calculate the speed of this second part.

21 Two space probes make a docking manoeuvre (join together) in space. One probe has a mass of 3000 kg and is travelling at 7·5 m s⁻¹. The second probe has a mass of 2000 kg and is moving at 6·0 m s⁻¹ in the same direction as the first space probe. Determine the common velocity of both probes after docking.

22 A squash ball, mass 25 g, collides with a wall at $66\,\text{m s}^{-1}$ and rebounds straight back in the direction from which it came at $48\,\text{m s}^{-1}$. Calculate the change in momentum Δp of the squash ball.

> **Hint** Remember Δp = final momentum – initial momentum.

23 A linear air track vehicle of mass 1·6 kg is travelling with a speed of $0\cdot5\,\text{m s}^{-1}$. It collides head-on with another vehicle of mass 1·2 kg, travelling at $0\cdot7\,\text{m s}^{-1}$ in the opposite direction. The vehicles lock together on impact. Determine the speed and direction of the vehicles after the collision.

24 Two air track vehicles with repelling magnets are held together by a light thread. Initially both vehicles are at rest and in contact with each other. The thread is carefully burned through and both vehicles move apart. One vehicle, with a mass of 2·00 kg, moves off with a speed of $2\cdot00\,\text{m s}^{-1}$. The other moves off with a speed of $1\cdot20\,\text{m s}^{-1}$ in the opposite direction. Calculate the mass of this vehicle.

25 Trolley 'X' has a mass of 2·8 kg and is travelling from left to right at $2\cdot4\,\text{m s}^{-1}$.

Trolley 'Y' has a mass of 3·4 kg and is travelling from right to left at $1\cdot6\,\text{m s}^{-1}$ as shown. When the two trolleys collide they stick together and move off as one. Find the velocity of both trolleys immediately after the collision.

26 A spring-loaded launch pad is mounted vertically on the top of a trolley. A ball is then placed onto the loaded compressed spring. The mass of the ball is 20 g and the mass of the trolley is 0·5 kg.

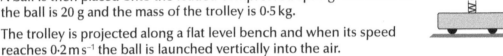

The trolley is projected along a flat level bench and when its speed reaches $0\cdot2\,\text{m s}^{-1}$ the ball is launched vertically into the air.

a What is the speed of the trolley immediately after the ball is released?

b Describe the motion of the ball until it lands.

27 A firework is launched vertically and when it reaches its maximum height it explodes into two pieces.

One piece has a mass of 250 g and moves off with a speed of $8\,\text{m s}^{-1}$. The other piece has a mass of 180 g.

What is the velocity of the smaller piece of the firework immediately after separation?

28 Two explosive trolleys are placed in contact with each other at rest on a bench.

explosive trolleys

Trolley 'A' has a mass of 1·5 kg and trolley 'B' has a mass of 0·5 kg. When the trolleys are allowed to explode apart they move off in opposite directions to each other. If the velocity of trolley 'A' is $2\cdot0\,\text{m s}^{-1}$ immediately after the explosion, what is the velocity of trolley 'B'?

29 Two air track vehicles travel towards one another and interact head-on as shown.

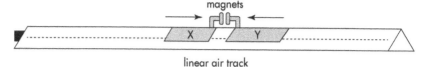

linear air track

The magnets are used to repel the two vehicles as they approach each other.

Mass of vehicle Y (kg)	Speed of vehicle X before interaction (m s⁻¹)	Speed of vehicle Y before interaction (m s⁻¹)	Speed of vehicle X after interaction (m s⁻¹)	Speed of vehicle Y after interaction (m s⁻¹)
0·20	1·44	1·20	1·84	1·44

Find the mass of vehicle X.

30 Two lab vehicles collide with each other and stick together. Vehicle '1', mass 2·4 kg, is travelling from left to right at 1·2 m s⁻¹ and vehicle '2' is travelling from right to left at 3·4 m s⁻¹. If the velocity immediately after the collision is 0·5 m s⁻¹ to the left, what is the mass of vehicle '1'?

> **Hint** Watch the direction signs!

31 A skateboard, mass 3·2 kg, is travelling at 2·2 m s⁻¹ horizontally when a student, mass 62 kg jumps vertically downward onto the skateboard from above. Find the speed of the student and skateboard immediately after the student makes contact.

> **Hint** The student has no horizontal velocity.

32 The following apparatus was set up to find the speed of an air-rifle pellet.

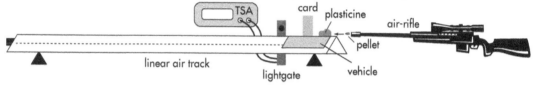

The pellet was fired into a lump of plasticine mounted on a stationary linear air track vehicle. On impact the vehicle and card move off and immediately pass through a lightgate that is attached to a Time/Speed/Acceleration (TSA) computer to measure the speed of the card and hence the vehicle. The following data was recorded:

Mass of vehicle, card and plasticine (kg)	Mass of 100 air-rifle pellets (kg)	Time for card to pass through lightgate (s)	Width of the card (m)
0·96	0·12	0·32	0·04

a Using the data provided, find the speed of the vehicle immediately after collision.

b Describe step-by-step how the speed of the pellet can be determined using this apparatus.

> **Hint** Use conservation of momentum.

c Find the speed of the pellet using the data provided.

d Why was the lightgate placed very close to the card?

 33 **a** Define what is meant by (i) an elastic collision, and (ii) an inelastic collision.

b Decide whether the collision in question 11 is elastic or inelastic.

34 A linear air track vehicle 'A' is projected towards another vehicle 'B' that is stationary.

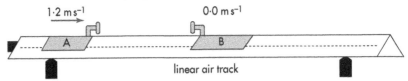

1·2 m s⁻¹   0·0 m s⁻¹

linear air track

Vehicle 'A' has a mass of 0·5 kg and is travelling at 1·2 m s⁻¹ to the right as shown. After colliding both vehicles move off separately to the right. Vehicle 'A' moves with a speed of 0·2 m s⁻¹ and vehicle 'B' moves with a speed of 0·5 m s⁻¹.

a What is the mass of vehicle 'B'?

b Is kinetic energy conserved?

c What type of collision is it?

d It is decided to repeat the experiment several times. Describe a simple way in which vehicle 'A' could always be projected with the same velocity each time.

35 A linear air track is used to measure the speed of an air-rifle pellet.

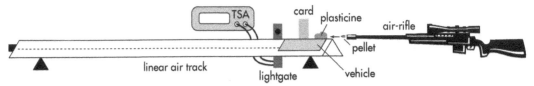

TSA   card   plasticine   air-rifle
pellet
linear air track   lightgate   vehicle

The actual speed of the pellet is calculated using the following equation **after** the plasticine is struck by the pellet:

$$speed\ of\ pellet = \frac{final\ mass\ of\ vehicle \times speed\ of\ vehicle}{mass\ of\ pellet}$$

Data captured from one experiment:

Final mass of vehicle (including plasticine and pellet) = $(1 \cdot 40 \pm 0 \cdot 02)$ kg

Mass of pellet = $(1 \cdot 17 \pm 0 \cdot 01)$ g

Speed of vehicle after being struck (including plasticine and pellet) = $(0 \cdot 110 \pm 0 \cdot 001)$ m s⁻¹

Provide a good estimate for the speed of the pellet (include an uncertainty).

36 Two linear air track vehicles 'A' and 'B' are connected by a stretched elastic band as shown.

elastic band

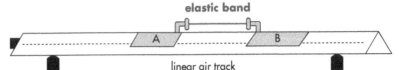

A   B

linear air track

Vehicle 'A' has a mass of 1·0 kg and vehicle 'B' has a mass of 500 g. The vehicles are carefully pulled apart, stretching the elastic band and then released. If the speed of vehicle 'A' immediately before impact was 0·5 m s⁻¹, what would be the speed of vehicle 'B' immediately before impact?

> **Hint** Use conservation of momentum.

37
A vehicle and small motorised fan have a mass 'm'. The vehicle rests on a plank of wood, mass $3m$, which is supported by a cushion of air from below.

When the fan motor is switched on the vehicle moves to the left. When the vehicle reaches a velocity 'v' what will be the velocity of the plank of wood?

> **Hint** Use conservation of momentum.

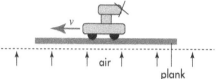

Exercise 3B Impulse

> **Hint** Assume friction is negligible throughout.

Example

A student investigates the force exerted by a snooker cue on a stationary snooker ball.

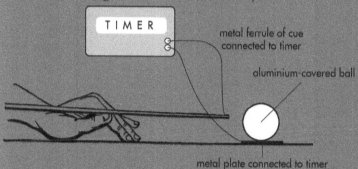

With the cue tip removed, the metal ferrule makes contact with the aluminium-covered ball acting as a switch, allowing the timer to measure the contact time between the two. The speed of the ball after it leaves the cue is measured using a lightgate connected to a computer.

DATA:

- Mass of ball = 0·17 kg
- Time of contact between the cue and ball = 0·005 s
- Speed of ball immediately after contact = 0·6 m s⁻¹

Calculate the average force exerted by the snooker cue on the snooker ball.

- First, write out the appropriate equation: $Ft = m(v - u)$
- Next, insert the data into the correct place $F \times 0.005 = 0.17 (0.6 - 0)$
- Finally, solve the problem to find F $\quad\quad F = \underline{20.4\ N}$

1 Explain why in sports like cricket, the player is taught to follow through with the ball when catching it rather than trying to catch it without following through.

2 A golf club exerts a maximum force of 9 kN on a golf ball for 0·50 ms.

Calculate the impulse on the ball.

3 A pool cue exerts an average force of 8·0 N on a stationary ball of mass 170 g. The impact of the cue on the ball lasts for 45·0 ms. What is the speed of the ball as it leaves the cue?

4 A stationary snooker ball, mass 170 g, is struck by a cue providing it with an initial speed of 1·8 ms⁻¹. The time of contact between the cue and the ball is 45 ms. Find the average force exerted by the cue on the ball.

5 A golf ball, mass 45 g, is struck by a golf club and takes off with a speed of 66 ms⁻¹. The average force exerted by the golf club on the ball is 2·2 kN. Calculate the time of contact between the club and the ball.

6 A student kicks a stationary football, mass 440 g, in a straight line towards the goalmouth. The student's foot is in contact with the ball for a time of 40 ms and the ball takes off with a velocity of 10 ms⁻¹. Find the average force exerted by the student's boot on the football.

7 A golf ball of mass 46 g falls from a height of 0·45 m onto concrete. The ball rebounds to a height of 0·28 m. The duration of the impact is 25 ms.

Calculate:

a the change in momentum of the ball when it bounces;

b the impulse on the ball during the bounce; and

c the unbalanced force exerted on the ball during the bounce.

8 A projectile is fired horizontally from a linear air track vehicle which is initially at rest.

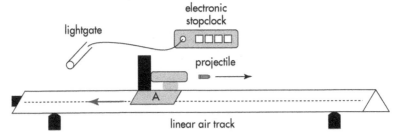

The electronic stopclock starts counting when the lightgate assembly is interrupted by the card attached to the air track vehicle.

The following data is collected:

• Mass of air track vehicle after launch of projectile = 0·45 kg

• Width of card = 0·04 m

• Time for card to pass through the lightgate = 0·28 s

Calculate the impulse imparted to the vehicle when the projectile is fired.

9 A rubber ball of mass 45·0 g is dropped from a height of 1·80 m onto a concrete floor. The ball rebounds to a height of 0·65 m. The average force of contact between the concrete and the ball is 3·75 N.

  a Calculate the velocity of the ball just before it hits the ground.

  b Calculate the velocity of the ball just after hitting the ground.

  c Calculate the time of contact between the ball and the concrete.

10 A golfer strikes a golf ball, mass 45·0 g, high into the air. During the strike the golf club exerts an average force of 4·0 kN and spends 0·55 ms in contact with the ball. Find the speed of the ball immediately after impact.

11 A catapult elastic is in contact with a stone, mass 55 g, for a time of 25 ms during firing. The stone leaves the catapult with a speed of 7·5 m s^{-1}.

  a Calculate the average force exerted by the catapult elastic.

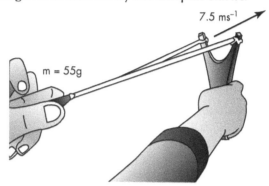

  b Sketch a graph to show how the force varies with time from the instant the stone is released. Include numerical values on both axes.

12 A football, mass 440 g, is dropped from a height of 1·60 m onto a concrete surface. It rebounds to a height of 0·92 m. The ball is in contact with the ground for 25 ms.

  Calculate:

  a the change of momentum of the ball during the bounce; and

  > **Hint** Work out E_p and equate to E_k to find v before and after the bounce.

  b the average force exerted by the surface on the ball.

13 A knowledge of 'impulse' is very important when car manufacturers are considering the safety features of new cars. Describe **two** design features in modern cars that make use of impulse to improve safety.

14 During a game of football a stationary ball of mass 0·42 kg is struck by a player during a penalty shoot-out. The graph shows how the force of the strike varied with time.

  a Calculate the speed with which the ball takes off after being struck.

  The next day during indoor training a softer ball of the same mass is used. During a practice penalty shoot-out this softer ball moves off with the same speed as the ball in part (a).

  b Sketch a graph to show how the force on the softer ball varies with time. Indicate how your graph compares to the one provided above.

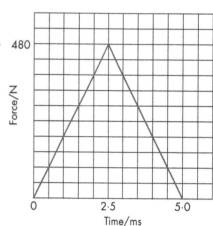

15 The graph below shows how the force exerted on an ice-hockey puck varies with time.

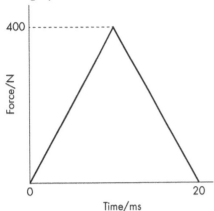

a The puck has a mass of 170 g. Assuming the puck was stationary to start with, calculate the speed with which it leaves the hockey stick.

> **Hint** Remember impulse = change of momentum.

b Assuming the change of momentum remains the same, re-draw the graph to show what would happen if the maximum force exerted was only 200 N. Include values on both axes.

16 A ball of mass 500 g falls from rest and hits the ground. The velocity–time graph shown represents the motion of the ball for the first 1·2 seconds after it starts to fall.

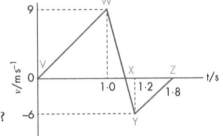

a Describe the motion of the ball during sections VW, WX, XY and YZ in the graph.

b How long was the ball in contact with the ground?

c Calculate the average unbalanced force that the ground exerts on the ball.

d How much energy is lost by the ball due to contact with the ground?

e What happens to this energy?

17  A ballistic pendulum can be used to determine the velocity of a bullet. The pendulum consists of a wooden block suspended by two lightweight nylon wires.

As the bullet enters the wooden block, momentum from the bullet is transferred to the block, causing it to rise as shown.

During one test the following data is recorded:

- Mass of wood block = 4·0 kg
- Mass of bullet = 4·2 g
- Height risen by pendulum = 0·05 m

Air resistance can be ignored throughout.

a Calculate the horizontal velocity of the pendulum immediately after it is struck by the bullet.

> **Hint** Consider conservation of energy.

b Calculate the horizontal velocity of the bullet immediately before it strikes the wooden block.

c Find the average force exerted on the wooden block by the bullet if the impact lasts for 2 ms.

d Describe a safe method to measure the height risen by the block after it is struck.

18 Baking soda crystals are used to sandblast a wall to remove graffiti. A mass of 1000 kg of crystals strikes the wall each minute with an average speed of 30 m s⁻¹. Upon impact with the wall the crystals fall vertically towards the ground. Calculate the average force exerted by the crystals upon the wall during the sandblasting.

> **Hint** Calculate the change in momentum per second.

19 A rocket engine ejects gas at a rate of 65 kg per second. The ejected gas has a constant speed of 1 750 m s⁻¹. Calculate the magnitude of the force exerted by the ejected gas on the rocket.

4 Projectile motion

Exercise 4A Projectiles

> Air resistance may be assumed to be negligible throughout.

Example

A football is kicked with a velocity of $20\,\text{m s}^{-1}$ at an angle of $30°$ to the horizontal on a flat playing field.

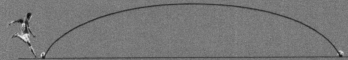

a Calculate the horizontal component of velocity of the ball.

b Calculate the vertical component of velocity of the ball.

c Calculate how far horizontally the ball travels until it first strikes the ground.

- First, use the appropriate equation for the horizontal velocity:

$$v_H = v \cos \theta$$
$$= 20 \cos 30°$$
$$= \underline{17 \cdot 3\,\text{m s}^{-1}}$$

- Next, use the appropriate equation for the vertical velocity:

$$v_v = v \sin \theta$$
$$= 20 \sin 30°$$
$$= \underline{10 \cdot 0\,\text{m s}^{-1}}$$

- Next, calculate the time of flight in the vertical plane to the highest point where the vertical velocity is temporarily zero using an equation of motion and then double this time to find the total time of flight:

$$v_v = u_v + gt$$
$$0 = 10 \cdot 0 + (-9 \cdot 8)t$$
$$t = 1 \cdot 02\,\text{s}$$
$$\text{total time of flight} = 2 \cdot 04\,\text{s}$$

- Finally, calculate the distance travelled horizontally using this time of flight.

$$v_H = \frac{d}{t}$$
$$17 \cdot 3 = \frac{d}{2 \cdot 04}$$
$$d = \underline{35 \cdot 3\,\text{m}}$$

1 A shuttlecock is launched horizontally over a net at 25 m s^{-1} and lands 5·4 m from the point it was struck.

  a Describe and explain the path taken by the shuttlecock.

  b Find the time of flight of the shuttlecock.

  c Find the vertical speed when the shuttlecock reaches the ground.

  d Find the velocity of the shuttlecock when it reaches the ground.

  e Sketch a graph of (i) the horizontal speed against time and (ii) the vertical speed against time for the shuttlecock. Include numerical values on both sets of axes.

2 A stunt motorcyclist takes off from a pier horizontally and lands on a small barge 15·0 m from the edge of the pier.

The height of the pier above the landing point is 5·50 m.

  a How long does the motorcyclist take to reach the barge after leaving the pier?

  b Calculate the minimum horizontal speed of the motorcyclist required in order to reach the barge.

3 A golf ball is dropped by a golfer and bounces vertically upwards off a concrete surface. The height reached by the ball is 0·360 m. Calculate the initial vertical velocity of the ball as it rises from the concrete.

4 An archer fires an arrow horizontally at 110 m s^{-1} at a target situated 50·0 m away. The arrow leaves the bow at a vertical height of 1·6 m. The top of the target is 1·2 m above the ground and the bottom of the target is 0·40 m above the ground.

Calculate where the arrow hits the target.

5 A plane is travelling with a constant horizontal velocity of 400 m s^{-1} at a height of 250 m. A box of emergency supplies is dropped vertically from the plane.

  a Calculate the time taken for the box to reach the ground.

  b Calculate the horizontal distance between the point where the box is released and the point where it lands.

  c How far horizontally is the plane from the box when the box hits the ground, assuming the plane continues to fly at constant velocity?

  d How far vertically is the plane from the box when the box hits the ground, assuming the plane continues to fly at constant velocity?

6 A football of mass 400 g is thrown from a height of 2·4 m with a velocity of 12 m s^{-1} at an angle of 35° above the horizontal.

Calculate:

  a the horizontal component of the initial velocity of the ball;

  b the vertical component of the initial velocity of ball;

  c the maximum vertical height above the ground reached by the ball;

  d the time of flight for the whole trajectory; and

  e the horizontal range of the ball when it lands on the ground.

7 A long jumper has a vertical component of velocity at take-off of $5.42\,\text{m}\,\text{s}^{-1}$. The horizontal distance travelled by the jumper is $8.52\,\text{m}$.

    a Calculate the horizontal velocity on take-off.

    b Calculate the angle to the horizontal on take-off.

8 In each of the following situations find (i) the horizontal component of velocity and (ii) the vertical component of velocity of the projectile.

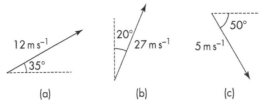

    (a)   (b)   (c)

9 A cannon inclined at $20°$ to the horizontal fires a cannonball at $120\,\text{m}\,\text{s}^{-1}$ towards a castle wall situated $914\,\text{m}$ away.

    a Calculate (i) the horizontal component of velocity and (ii) the vertical component of velocity of the cannonball.

    b How long does the cannonball take to reach its maximum height?

    c What is the maximum height reached by the cannonball?

    d How long is the cannonball in the air?

    e Assuming the base of the castle wall is at the same level as the cannon, how high up the wall does the cannonball strike the wall?

10 A cannonball is fired horizontally with a speed of $55.0\,\text{m}\,\text{s}^{-1}$ at a height of $1.50\,\text{m}$ above the edge of a cliff. The cannonball strikes the sea below at a point $140.0\,\text{m}$ from the base of the cliff.

    a Calculate the time of flight of the cannonball.

    b Calculate the velocity of the cannonball $2.0\,\text{s}$ after it is fired.

    c What is the height of the cliff above sea level?

11 An archer fires an arrow at a target $75\,\text{m}$ away. The arrow is fired at $33\,\text{m}\,\text{s}^{-1}$ at an angle of $28°$ as shown.

    a What is the vertical component of the velocity of the arrow?

    b What is the horizontal component of velocity of the arrow?

    c How long will it take for the arrow to hit the bullseye if the arrow takes off level with the bullseye?

    d Why does the archer not fire *directly* at the bullseye?

12 A rugby player is taking a penalty kick. The player kicks the ball at an angle of 32° to the horizontal with an initial speed of 16 m s⁻¹.

The goal posts are 22 m away directly in front of the ball. To score, the player must kick the ball over the cross bar, which is 3·0 m above the ground. Does the player score? Justify your answer.

13 A golf ball is struck with a velocity, v, at an angle θ to the horizontal. A tree, height 35 m, lies exactly halfway between the point where the ball is struck and the point where the ball first lands.

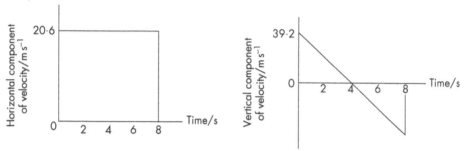

Not to scale

Graphs of the horizontal and vertical motion are shown below.

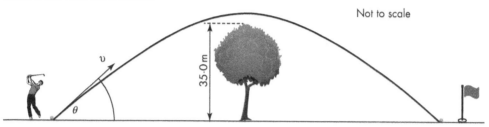

a How far above the tree is the ball when it passes overhead?

b How far horizontally does the ball travel between being struck and hitting the ground?

c Find the initial velocity of the ball when it is struck.

14 A ball is thrown horizontally with a speed of 25 m s⁻¹ from a cliff. How long after being thrown will the velocity of the ball be at an angle of 40° to the horizontal?

15 A ball released from rest at a height 'h' takes 4·0 seconds to reach the ground on Earth. On another planet the gravitational field strength is 2·7 times greater than that on Earth.

Calculate how long it will take the ball to drop from rest through the same height assuming air resistance is negligible on both planets.

16 An aeroplane flies with a constant horizontal velocity of 120 m s⁻¹. It releases a package that is required to drop into a clearing in the jungle from a height of 90 m.

package

Calculate how far in front of the clearing the plane must be when it releases the package in order to deliver the load to the precise location.

17 A toy car rolls down a slope and off the end of a 1·28 m high bench as shown opposite.

The car leaves the slope at an angle of 45° with a speed of 1·5 m s⁻¹.

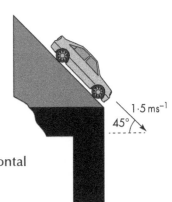

1·5 ms⁻¹

45°

a Calculate how long it takes the car to reach the floor.

b How far horizontally from the base of the bench does the car land?

c The angle of slope is now decreased. How does the horizontal distance where the car lands compare to that in part (b)? Justify your answer.

18 A golfer needs to get his golf ball out of a bunker and onto the green as shown.

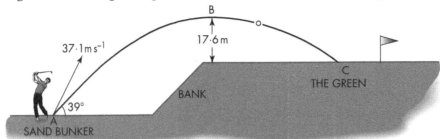

B

37·1 m s⁻¹

17·6 m

o

C

THE GREEN

BANK

39°

A

SAND BUNKER

The golf ball is initially struck at point 'A' with a velocity of 37·1 m s⁻¹ at an angle of 39°. The ball reaches its maximum height at point 'B' 17·6 m above the green. The ball first strikes the green at point 'C'.

a Find (i) the horizontal component of velocity and (ii) the vertical component of velocity of the golf ball when it is struck.

b Find the time taken for the ball to reach the green at point 'C'.

> Hint | Break the flight of the ball into two halves.

c Calculate how far horizontally the ball travels between points 'A' and 'C'.

19 A stone is thrown horizontally from the top of a very high dam. Assuming no air resistance, is there any point during its fall that the stone travels vertically only? Justify your answer.

20 A soldier sitting in the back of a jeep travelling at 12 m s⁻¹ horizontally accidently fires a bullet vertically upwards when the jeep hits a stone in the road. Assuming the jeep maintains this constant velocity, is the soldier in any danger when the bullet returns to the height from which it was fired? You must justify your answer.

21 A projectile is fired at an angle of 40° with an initial speed 'u' to the horizontal as shown.

If the projectile reaches its highest point after a time, 't', provide an expression for its horizontal range using u, t and the angle 40°.

> **Hint** Consider the only equation used in the horizontal plane.

22 A basketball player throws a ball with an initial velocity of 6·30 m s⁻¹ at an angle of 40° to the horizontal. The ball travels a horizontal distance of 2·66 m to reach the hoop supporting the basket. The height of the hoop is 3·05 m above the court.

a Find (i) the horizontal component of velocity and (ii) the vertical component of velocity of the ball when it is struck.

> **Hint** Work out first the distance travelled vertically in the time the ball takes to reach the hoop.

b Find how long it takes for the ball to reach the hoop.

c Find the height above the court from which the ball was released.

23 A tennis ball is struck from a height of 2·60 m with a service speed of 42·8 m s⁻¹ at an angle of 5° below the horizontal. The net is 6·40 m away from the service line and the height of the net is 0·91 m. Determine how far above the net the ball is when it passes over the net.

> **Hint** Work out the time to reach the net first.

5 Gravitation

Exercise 5A Gravitation

> **Gravitational constant: $6 \cdot 67 \times 10^{-11}\ \mathrm{N\,m^2\,kg^{-2}}$**

Example

Two planets have masses of $7 \cdot 22 \times 10^{24}$ kg and $3 \cdot 34 \times 10^{26}$ kg. The two planets are $54 \cdot 1 \times 10^6$ m apart. Calculate the gravitational force of attraction between the two planets.

Use the appropriate equation $\quad F = \dfrac{Gm_1 m_2}{r^2}$

$$F = \frac{\left(6 \cdot 67 \times 10^{-11} \times 7 \cdot 22 \times 10^{24} \times 3 \cdot 34 \times 10^{26}\right)}{\left(54 \cdot 1 \times 10^6\right)^2}$$

$$F = \underline{5 \cdot 50 \times 10^{25}\ \mathrm{N}}$$

1 Find the missing values in each case:

Mass of Sun: $1 \cdot 99 \times 10^{30}$ kg

Planet	Mass (kg)	Mean distance from the Sun ($\times 10^9$ m)	Force between Sun and the planet (N)
Mercury	$3 \cdot 29 \times 10^{23}$	57·9	
Venus		108·2	$5 \cdot 52 \times 10^{28}$
Earth	$5 \cdot 97 \times 10^{24}$		$3 \cdot 54 \times 10^{28}$
Mars	$6 \cdot 34 \times 10^{23}$	227·9	
Jupiter	$1 \cdot 90 \times 10^{27}$		$4 \cdot 16 \times 10^{29}$
Saturn		1433·5	$3 \cdot 67 \times 10^{28}$
Uranus	$8 \cdot 68 \times 10^{25}$		$1 \cdot 40 \times 10^{27}$
Neptune	$1 \cdot 02 \times 10^{26}$	4495·1	

2 Two students sit 2·0 m apart in an examination room. One student has a mass of 45 kg and the other has a mass of 55 kg. Find the gravitational force of attraction between the two students.

3 Find the gravitational force of attraction between two 5·0 kg masses that are 22 cm apart.

4 Calculate the gravitational force of attraction between a man weighing 784 N standing on the Earth's equator. The mean equatorial radius of the Earth is $6 \cdot 378 \times 10^6$ m and its mass is $5 \cdot 97 \times 10^{24}$ kg.

5 Show that $\mathrm{N\,m^2\,kg^{-2}}$ are appropriate units for the gravitational constant, G.

6 On a particular day the International Space Station (ISS) is situated 400 km above the Earth's surface at the Equator. The mass of the Earth is $5 \cdot 97 \times 10^{24}$ kg and the mass of the ISS is $3 \cdot 70 \times 10^5$ kg. The mean equatorial radius of the Earth is $6 \cdot 378 \times 10^6$ m.

Find the gravitational force of attraction between the Earth and the ISS at this point.

7 In a hydrogen atom, an electron, mass $9 \cdot 11 \times 10^{-31}$ kg, is in orbit around a proton, mass $1 \cdot 673 \times 10^{-27}$ kg, at a mean distance of $5 \cdot 29 \times 10^{-11}$ m.

Find the gravitational force of attraction between the proton and the electron.

8 Two protons are $4 \cdot 00 \times 10^{-15}$ m apart and exert a gravitational force on each other. The mass of a proton is $1 \cdot 67 \times 10^{-27}$ kg. Calculate the size of this gravitational force.

9 Two objects are situated at a distance of $1 \cdot 25$ m between their centres. The gravitational force acting between the two of them is $9 \cdot 32 \times 10^{-11}$ N. If the mass of one of them is 520 g, find the mass of the other one.

10 Neptune (mass $1 \cdot 024 \times 10^{26}$ kg) is currently thought to have 14 moons in orbit around it. One of these moons is Proteus (mass $4 \cdot 40 \times 10^{19}$ kg). The gravitational force between Neptune and Proteus at a particular instant in time is $2 \cdot 15 \times 10^{19}$ N.

Calculate the approximate distance in km between the centre of Proteus and the centre of Neptune at this instant in time.

11 Two satellites are in orbit around a distant planet. One satellite has a mass of $1 \cdot 22 \times 10^{3}$ kg and the other has a mass of $7 \cdot 35 \times 10^{4}$ kg. The gravitational force of attraction between the two satellites is $5 \cdot 75 \times 10^{-13}$ N.

Calculate the distance in km between the two satellites.

12 Alpha Centauri A and Alpha Centauri B form the nearest binary star system at a mean distance of $4 \cdot 221$ light years from Earth. The two stars orbit a common centre of gravity with an average radius of approximately $8 \cdot 230 \times 10^{8}$ km. The mass of Alpha Centauri A is $2 \cdot 188 \times 10^{30}$ kg. If the gravitational force of attraction between these two stars is $9 \cdot 72 \times 10^{25}$ N, calculate the approximate mass of Alpha Centauri B to 4 significant figures.

13 A planet orbits a star at a distance of $5 \cdot 6 \times 10^{6}$ km. The star exerts a gravitational force of $2 \cdot 6 \times 10^{27}$ N on the planet. The mass of the star is $8 \cdot 5 \times 10^{31}$ kg. Calculate the mass of the planet.

14 A satellite is in space at a height of $1 \cdot 2 \times 10^{2}$ km above the Earth's equator. The mass of the Earth is $5 \cdot 97 \times 10^{24}$ kg and its mean equatorial diameter is $12 \cdot 76 \times 10^{6}$ m. The gravitational force of the Earth on the satellite is $2 \cdot 15 \times 10^{3}$ N.

Find the weight of the satellite on the Earth's surface.

15 What is the difference between 'g' and 'G'?

16 Every so often Mercury, Venus and the Earth are perfectly aligned as illustrated below.

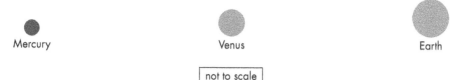

Mercury   Venus   Earth

not to scale

Using the information in the table below, calculate the **resultant** gravitational force on Venus from Mercury and the Earth when the three celestial bodies are aligned in a straight line as shown.

Celestial body	Mean distance from the Sun ($\times 10^{6}$ km)	Mass (kg)
Mercury	57·9	$0 \cdot 33 \times 10^{24}$
Venus	108·2	$4 \cdot 87 \times 10^{24}$
Earth	149·6	$5 \cdot 97 \times 10^{24}$

Ignore the gravitational influence of any other celestial bodies.

17 The mean radius of the Earth at the equator is 6.378×10^3 km and the mass of the Earth is 5.97×10^{24} kg.

   a Find the gravitational force of the Earth on a mass of 5000 kg at:

     i the equatorial surface;

     ii twice the equatorial radius;

     iii four times the equatorial radius; and

     iv eight times the equatorial radius.

   b Express the force in each case as a fraction of the force at the equatorial radius.

   c Do the values calculated in part (b) support the inverse square law for gravitation?

18 Three objects X, Y and Z have masses of 0·6 kg, 0·9 kg and 1·4 kg respectively. They are suspended vertically by very light nylon threads (not shown) and all lie in a straight line separated by the distances shown.

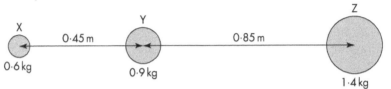

   a Find the resultant gravitational force acting on Z.

> **Hint** Find the force between X and Z and add to the force between Y and Z.

   b Calculate the distance **from X** along the straight line that Z must be moved so that there is no resultant gravitational force acting on Y.

> **Hint** Find the force between X and Y and then equate this to the force between Y and Z to find the new distance of Z from Y.

19 The Earth has a mass of 5.97×10^{24} kg and the Moon has a mass of 7.36×10^{22} kg. The mean orbital radius of the Moon as it orbits the Earth is 3.80×10^5 km. How far from the Earth will the gravitational attraction to the Moon be equal and opposite to the gravitational attraction to the Earth?

> **Hint** Equate $\dfrac{GM_E m}{r^2}$ to $\dfrac{GM_m m}{r^2}$ for a mass 'm' and find the ratio of the distance to Earth/distance to the Moon.

20 Explain why the Earth has two high tides each day on opposite sides of the Earth as the result of the interaction between the Moon and the Earth.

6 Special relativity

Exercise 6A Special relativity

1 The term 'frame of reference' is often used in relativity.

  a What does the term 'frame of reference' mean?

  b Provide two examples of a frame of reference.

2 Albert Einstein is inside a closed lift that is descending at a constant velocity. Einstein decides to conduct an experiment involving the measurement of the time taken for a ball to drop from a height of 1·8 m to the floor of the lift. His colleague Niels Bohr carries out the exact same experiment with the same equipment in a stationary laboratory.

  a Do Einstein and Bohr record the same time for the ball to fall through 1·8 m?

  b Is it possible from the experiment alone for Einstein to determine that he is moving with constant velocity?

  c What does the (correct) answer to part (b) tell us about the physical laws governing special relativity with regard to the frame of reference used?

3 In the equation for special relativity $t = \dfrac{t'}{\sqrt{1 - \dfrac{v^2}{c^2}}}$ what does:

  a t represent?

  > **Hint** Remember that $t < t'$.

  b t' represent?

  c v represent?

4 What is meant by the term 'time dilation' often used in special relativity?

5 What happens to the time, t, in the above equation when an object gets closer and closer to the speed of light, c?

6 The relativistic effects of time dilation are not observed in everyday life. Why is this?

7 Is it possible to accelerate a proton to the speed of light? Justify your answer.

Example

A subatomic particle has a mean lifetime of $1·26 \times 10^{-5}$ s in its own frame of reference. The particle is travelling past the Earth at $0·992c$.

a Calculate the distance travelled by the subatomic particle in its own frame of reference.

Use the appropriate equation: $d = vt = (0·992 \times 3·0 \times 10^8) \times 1·26 \times 10^{-5} = \underline{3·7 \times 10^3 \text{ m.}}$

b Calculate the mean lifetime of the subatomic particle according to an observer on Earth.

Use the appropriate equation: $t = \dfrac{t'}{\sqrt{1 - \dfrac{v^2}{c^2}}}$ $t = \dfrac{1·26 \times 10^{-5}}{\sqrt{1 - (0·992)^2}} = \underline{9·98 \times 10^{-5} \text{ s}}$

8 An experiment is carried out on a spaceship moving at $2.2 \times 10^8\,\mathrm{m\,s^{-1}}$ that involves the measurement of time by an astronaut on the spaceship between two events also on the spaceship. The time measured was 25 ns. What would be the time measured for the same event according to an observer on Earth?

9 Two astronauts on different spaceships are observing a 'light clock', which consists of a beam of light bouncing vertically between two mirrors that are 3.0 m apart. The light clock is mounted on a moving spaceship that has a velocity of $0.5c$ relative to the other (stationary) spaceship.

What is the time taken for the light beam to pass from one mirror to the other mirror according to (i) the astronaut on board the spaceship that is moving and (ii) the astronaut on board the stationary spaceship?

> **Hint** First establish which is t and t'.

10 A particle passes between two detectors A and B situated 32.2 m apart. The time taken for the particle to pass between the two detectors is 137 ns.

a Find the speed of the particle according to a stationary observer next to the detectors.

b Calculate the proper time for the particle to pass between the two detectors.

11 A muon is a subatomic particle produced when a pion decays after the interaction between two protons. Stationary muons have a half-life of 1.9 μs. A muon is moving with a speed of $0.83c$ relative to a laboratory frame of reference. Calculate the half-life of the muon measured by the laboratory clock.

12 A technician uses a light clock to measure the return time for a beam of light between two mirrors on a stationary spaceship. The technician measures a time of 8.0 ns. Another technician in a moving spaceship measures a time of 19.6 ns. Calculate the speed at which the spaceship is moving relative to the stationary spaceship.

13 As a spaceship passes from one moon to another at a constant speed, a stopclock on the planet below and a stopclock on the spaceship are started simultaneously.

a An observer on the planet with the stopclock records a time of 7 hours for the journey and simultaneously records a time of 6 hours from a stopclock on the spaceship. What is the speed of the spacecraft according to this observer?

b What is the speed of the planet (with the stopclock) according to an observer on the spaceship?

c The observer on the spaceship can also see both stopclocks. When the spaceship clock reads 7 hours, what time does the stopclock on the planet read?

14 Muons are created high up in the Earth's atmosphere, travel at $0.9995c$ and decay with a half-life of 2.2 μs. A detector positioned on a mountain 2 km from the Earth's surface detects 2050 muons in a period of 300 s. An identical detector at the foot of the mountain detects 1370 muons in the same time period.

a Find the time taken for the muons to travel between the two detectors according to the muon's frame of reference.

b According to the muon's frame of reference, how many muons would be expected to arrive at the second detector?

c Explain the difference between the number actually detected and your answer to (b).

15 In the equation for special relativity; $l' = l\sqrt{1 - \dfrac{v^2}{c^2}}$ what does:

    a l represent?

    | Hint | Remember that $l' < l$. |

    b l' represent?

Example

An observer on the Moon measures the length of a rocket travelling at $0.85c$ to be $220\,m$ as it flies past. Calculate the length of the rocket according to an astronaut on board the rocket.

Use the appropriate equation: $l' = l\sqrt{1 - \dfrac{v^2}{c^2}}$ $220 = l \times \sqrt{1 - 0.85^2}$ $l = \underline{418\,m}$

16 An observer on Earth measures the length of a spaceship travelling at $0.8c$ to be 162 metres as it is flying past. Calculate the length of the spaceship according to an astronaut on board the spaceship.

    | Hint | First establish which is l and l'. |

17 The length of a rocket measured at rest on a launchpad is $111\,m$. The rocket takes off and eventually reaches a speed of $1.95 \times 10^8\,m\,s^{-1}$ as it passes the Moon. Calculate the length of the rocket according to an engineer on the Moon.

18 A beam of red light has a wavelength of $650\,nm$ measured by a technician in a stationary laboratory on Earth. When the same beam of light (on Earth) is observed by an astronaut on a moving spaceship the light appears more violet with a wavelength of $432\,nm$. Find how fast the spaceship is moving.

19 How fast would a $5\,m$-long car have to travel in order to fit into a $4\,m$-long parking space according to a stationary observer at the side of the parking space?

20 A spaceship travelling past a moon colony at $0.9c$ fires a beam of light forward from the nose of the spaceship. What speed does an observer on the moon see the beam of light travel at? Explain your answer.

7 The Doppler effect

Exercise 7A The Doppler effect

v_{sound} (air) = 340 m s^{-1} c (air/vacuum) = 3×10^8 m s^{-1}
Ignore any relative motion of the Earth.

1 A traveller standing on a railway bridge notices that the frequency of the whistle coming from a train appears to change when the train approaches, passes under and travels away from the bridge.

   a Describe the change that takes place in terms of the frequency of the whistle.

   b Explain in words and diagrams why the frequency of the note heard by the traveller standing on the bridge changes from the frequency emitted as the train approaches and travels away from the bridge.

2 A teacher rotates a battery-operated Doppler ball (siren) around her head at a constant speed as students stand a few metres away from the teacher. The siren emits sound waves with a frequency of 1150 Hz.

   a Describe what would be heard by the teacher swinging the ball in the circle. Explain.

   b What happens to the sound frequency heard by a student standing a few metres away as the ball travels towards the student. Explain.

   c What happens to the sound frequency heard by another student as the ball travels away from the student? Explain.

   d At what point in the rotation will a student hear the true frequency of the Doppler ball?

   e What change to the frequency shift (change in frequencies) would be observed if the ball was rotated much faster than before? Explain.

3 An ambulance uses a two-tone horn emitting frequencies of 960 Hz and 770 Hz. Calculate the frequencies heard by a pedestrian at the side of a road as the ambulance:

   a approaches at 15 m s^{-1}; and

   b recedes at 25 m s^{-1}.

Hint Remember to use '−' when approaching and '+' when moving away.

Example

A helicopter with a loudspeaker attached emits an 800 Hz signal to scare birds from an airfield. What will be the frequency heard by a stationary observer on the airfield as the helicopter travels at 80 m s^{-1} towards her?

- First, use an appropriate equation: $f_0 = f_s\left(\dfrac{v}{v - v_s}\right)$ – notice the '−' here because the helicopter is approaching!

- Insert the data in the appropriate place and solve $f_0 = 800\left(\dfrac{340}{340 - 80}\right) = \underline{1046\,Hz}$

4 An aircraft is flying at 150 m s^{-1} horizontally above a detector. The frequency of the sound it emits is 400 Hz. What will be the change in frequency as it flies overhead?

5 A police car is travelling on a straight, level part of a motorway towards an overhead bridge at 20 m s^{-1} using its two-tone horns. The frequencies heard by a pedestrian on the overhead bridge are 600 Hz and 861 Hz. Calculate the actual frequencies used by the police car's horns.

6 A 747 aircraft emits a sound with a frequency of 400 Hz when flying. Calculate the frequency heard by ground crew as it:

a approaches the runway at 82 m s^{-1}; and

b takes off from the runway at 70 m s^{-1}.

7 The frequency of a truck horn is 330 Hz when the truck is stationary. A truck is travelling on a straight level road towards an overhead bridge and the horn is sounded. The frequency heard by a pedestrian on the bridge is 362 Hz. Calculate the speed of the truck at this moment in time.

8 A bus is travelling along a straight level road when the driver sounds the bus horn. The bus horn frequency when stationary is 420 Hz but a student at a bus stop hears a frequency of 408 Hz.

a Is the bus heading towards or away from the bus stop?

b What is the speed of the bus at this moment in time?

9 A motorcyclist is travelling at a constant speed of 64 km h^{-1} along a straight level road towards a pedestrian waiting to cross the road. The frequency of sound heard by the motorcyclist at this speed when travelling towards the pedestrian is 64·2 Hz. What is the frequency of the sound heard by the pedestrian when the motorcyclist has passed and is now moving at 70 km h^{-1}?

> **Hint** First find f_s.

10 A truck's horn has a frequency of 440·0 Hz when the truck is stationary. A sound analyst on an overhead bridge measures the frequency of the sound from the horn as it is travelling at a constant speed towards the bridge. The frequency measured is 474·9 Hz. The truck accelerates slightly for 4 s as it passes under the bridge to a new constant speed as it moves away from the bridge. The new frequency measured at this constant speed is 407·6 Hz.

Calculate the acceleration of the truck.

> **Hint** Find u and v then apply an equation of motion.

11 A child is swinging back and forth on a swing. A stationary loudspeaker to the right of the child produces a sound with a fixed frequency, 'f'.

loudspeaker

> **Hint** Although the image shows a child moving towards a loudspeaker, the same frequency change is observed when the child is at rest and the speaker is moving.

    a Using the letters 'A', 'B' and 'C', identify where the child will hear a lower frequency than 'f'.

    b Using the letters, identify where the child will hear a higher frequency than 'f'.

    c At what point(s) will the highest frequency be heard?

    d At what point(s) will the lowest frequency be heard?

    e At what point(s) will the true frequency 'f' be heard?

12 When the Doppler effect is used with light from distant stars and galaxies the term 'redshift' is often used. What does this term mean?

13 What unit is used for the redshift, 'z', when considering the Doppler shift of light? Explain.

Example

A source of light emanating from a distant galaxy receding from Earth is observed to have a wavelength of 440×10^{-7} m. The same light detected from a stationary source on Earth has a wavelength of 436×10^{-7} m. How fast is this galaxy receding from the Milky Way?

- First, use an appropriate equation to get the redshift: $z = \dfrac{\lambda_{obs} - \lambda_{rest}}{\lambda_{rest}} = \dfrac{440 \times 10^{-7} - 436 \times 10^{-7}}{436 \times 10^{-7}}$

$$= 9.17 \times 10^{-3}$$

- Now use this value in the other redshift equation to find the speed: $z = \dfrac{v}{c}$

$$= 9.17 \times 10^{-3} = \frac{v}{3.0 \times 10^{8}}$$

$$v = \underline{\underline{2.75 \times 10^{6}\,\text{m s}^{-1}}}$$

14 A star is moving away from our solar system with a relative speed of 4.58×10^{5} m s^{-1}. What is the value of this star's redshift?

15 The same hydrogen emission spectrum from a stationary source in a laboratory and from a source moving at constant speed are shown below.

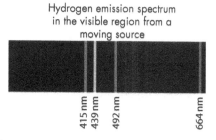

Hydrogen emission spectrum in the visible region from a stationary source

410 nm 434 nm 486 nm 656 nm

Hydrogen emission spectrum in the visible region from a moving source

415 nm 439 nm 492 nm 664 nm

What is the speed of the moving source?

16 Light from a distant galaxy contains the spectral lines of hydrogen. One of the lines has a measured wavelength of 656 nm. When the same line is observed from a hydrogen source on Earth it has a wavelength of 624 nm.

a Calculate the Doppler shift, z, for this galaxy.

b Calculate the speed at which this galaxy is moving relative to the Earth.

c Is this galaxy moving towards or away from the Earth? Explain.

> **Hint** Equate the two redshift formulae to each other.

17 A distant galaxy is travelling at a speed of 6.28×10^6 m s^{-1}. One particular wavelength of light in the hydrogen spectrum emitted by this galaxy has a value of 486 nm. What will be the wavelength of this line when the light reaches the Earth?

18 A distant galaxy has a redshift of 0.775. How fast is it moving away from the Earth?

19 The wavelength of a line in the hydrogen spectrum on Earth is 434.0 nm. The same line is observed from a star and has a wavelength of 438.2 nm.

a What is the star's redshift?

b How fast is the star moving away from the Earth?

20 A galaxy is moving away from the Earth at a speed of 2.80×10^7 m s^{-1}. One particular wavelength of light emanating from this galaxy is observed using a spectrometer attached to a telescope on Earth.

a Calculate the Doppler shift, z, for this galaxy.

b Calculate the wavelength of the light emanating from the galaxy using the spectrometer.

21 A galaxy is moving at 3.28×10^7 m s^{-1} away from the Earth. A particular wavelength of light emitted by this source is 553.4 nm. The same emission line is observed from a source on Earth and is found to have a frequency of 0.732×10^{15} Hz.

a What is the wavelength of the light from the stationary source on the Earth?

b Calculate the frequency of the light forming the emission line arriving from the galaxy.

22 A star is travelling away from the Earth at 4.52×10^6 m s^{-1}. The wavelength of a particular line in the emission spectrum of this star is 619.8 nm when observed from Earth. What is the actual wavelength of this particular light emitted by the star?

23 The wavelength of a particular line in the emission spectrum of an element in a distant star is 600·0 nm when it is observed from Earth. When the emission spectrum is produced in a laboratory on Earth the corresponding line has a wavelength of 589·6 nm.

a What is the redshift of this star?

b Calculate the velocity of the star.

> Hint First of all establish which wavelength is λ_o and which is λ_s.

24 A light source is moving at $5\cdot33 \times 10^6\,\text{m s}^{-1}$. The wavelength of a particular line in the emission spectrum of this light source is 553·4 nm. What is the actual frequency of this particular line when observed from Earth?

25 A distant galaxy is moving away from the Earth at a speed of $4\cdot32 \times 10^6\,\text{m s}^{-1}$. A light source emitted by this galaxy has a wavelength of 589·0 nm.

Calculate:

a the **shift** in wavelength; and

b the **shift** in frequency observed by an astronomer when this light source reaches the Earth.

> Hint 'Shift' means difference.

26 As the Sun rotates, light waves received from opposite sides of the Sun's equator show equal (in size) but opposite (in direction) Doppler shifts. The relative speed of rotation of a point on one side of the Sun (relative to the Earth) is $2\cdot07\,\text{km s}^{-1}$.

a Calculate the wavelength **shift** between the right-hand edge and the left-hand edge of the Sun's equator for a hydrogen line with a wavelength of 434 nm on Earth.

b What fundamental assumption have you made in calculating this wavelength shift?

8 Hubble's law and the Big Bang

Exercise 8A Hubble's law and the Big Bang

Hubble's constant $H_0 = 2.34 \times 10^{-18}\,\text{s}^{-1}$ Wien's constant $= 2.898 \times 10^{-3}\,\text{m K}$

1 What evidence is there for the 'Big Bang' theory?

2 What is Hubble's law?

3 Alpha Centauri is the nearest star system to our Sun and is situated 4·37 light years (ly) away. Calculate the distance to Alpha Centauri in metres.

4 One of the nearest galaxies to our Milky Way is called the Andromeda Galaxy, which is $2.4 \times 10^{22}\,\text{m}$ away from the Earth. Calculate how far this is in ly.

5 What is the velocity of the Sombrero Galaxy, which is 29.35×10^6 ly from Earth, according to Hubble's law?

6 What is the distance (in light years) to the Whirlpool Galaxy according to Hubble's law if its recessional velocity is $513\,\text{kms}^{-1}$?

7 Bode's Galaxy has a redshift of 8.71×10^{-4}.

  a How fast is it moving away from the Earth?

  b How far away from the Earth is it currently (in light years)?

8 The Sombrero Galaxy has a redshift of 5.7×10^{-3}.

  a How fast is this galaxy moving away from the Earth?

  b How far away from Earth is it currently in km?

9 **a** Starting with $v = H_0 d$ and applying the $d = vt$ relationship, derive an equation that allows the age of the universe to be found from Hubble's constant alone.

  b Using the equation derived in part (a), estimate the current age of the universe.

10 Astronomers can use the following relationship to determine the distance, d, to a star: $b = \dfrac{L}{4\pi d^2}$

The star Tau Ceti is spectrally similar to our own Sun and has the following physical properties:

- Apparent brightness, $b = 1 \cdot 27 \times 10^{-9} \, \text{W m}^{-2}$
- Luminosity, $L = 2 \cdot 03 \times 10^{26} \, \text{W}$

Based on this information, calculate the distance, d, to this star in ly.

11 What evidence is there for the existence of:

a dark matter; and

b dark energy?

12 Stellar physics makes reference to a type of radiation called 'black body radiation'. What is meant by a 'black body' and what is meant by 'black body radiation'?

13 The table below shows the values of λ_{peak} at different surface temperatures for a variety of stars.

Temperature (K)	λ_{peak} ($\times 10^{-6}$ m)
2000	1·45
4000	0·72
6000	0·48
8000	0·36
10000	0·29

Using this data, determine the relationship between temperature, T, and λ_{peak}.

Example

The peak wavelength emitted by a distant star is 555·0 nm. What is its surface temperature in kelvins?

- First, use an appropriate equation: $\lambda_{peak} T(\text{K}) = 2 \cdot 898 \times 10^{-3} \, \text{mK}$
- Next, insert data and solve, $555 \times 10^{-9} \times T(\text{K}) = 2 \cdot 898 \times 10^{-3}$

$$T = \underline{5222 \, \text{K}}$$

14 Sirius B has a surface temperature of 9940 K. What is its peak wavelength?

15 The peak wavelength of Epsilon Eridani is 570 nm. What is its surface temperature (in kelvins)?

16 Two stars, Rigel and Betelgeuse, form part of the constellation of Orion, easily visible throughout the year in the northern hemisphere. Rigel has a peak wavelength of 263 nm and Betelgeuse has a peak wavelength of 828 nm. Which of the two has a higher surface temperature and by how much?

17 Two stars, Castor and Pollux, have surface temperatures of 10300 K and 4865 K respectively. Which of these two stars has the shorter peak wavelength, and in which part of the electromagnetic spectrum is this wavelength to be found?

18 Edwin Hubble's original estimate in 1929 of the age of the universe was based upon a value of Hubble's constant, $H_0 = 1.62 \times 10^{-17}$ s^{-1}.

   a Calculate the age of the universe in years according to Hubble in 1929.

   b Why did this value of the age of the universe in 1929 raise concerns within the scientific community?

19 Why is it that, even though photons existed 300 000 years after the Big Bang, the universe was still opaque?

20 Cosmic microwave background radiation (CMBR) is part of the evidence for the Big Bang. What exactly is CMBR?

21 The temperature of the CMBR is 2·725 K. Explain why it is not possible to see this radiation with an optical telescope.

22 What is 'Olbers' paradox'?

23 The following graph shows how the intensity of radiation emitted by three different stars varies with the wavelengths of light emitted.

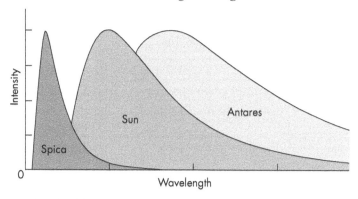

   a Which of the three stars is the hottest?

   b Which of the three stars is the coolest?

   c Which star emits more ultraviolet radiation than visible or infrared?

   d Spica has a photosphere temperature of 23 000 K. What is its peak wavelength, λ_{peak}?

   e Antares has a peak wavelength, λ_{peak}, of 852 nm.

     i What part of the e-m spectrum does this wavelength correspond to?

     ii Calculate the photosphere temperature of Antares in kelvin, K.

9 The Standard Model

Exercise 9A The Standard Model

1 Below is a representation of the Standard Model of fundamental particles. Using the item bank, choose the correct word for each part of the image.

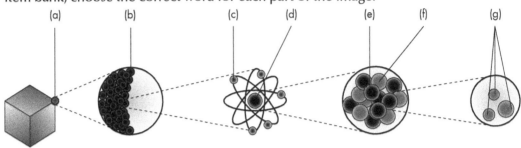

		ITEM BANK				
electron	nucleus	quarks	matter	proton	atom	neutron

2 The following list of objects is not in any particular order:

height of a human being	diameter of a typical nucleus	area of Russia
height of the Eiffel Tower	diameter of an electron	diameter of the Moon
diameter of a hydrogen atom	distance to nearest galaxy	diameter of Andromeda Galaxy
diameter of the Solar System	diameter of a neutron	height of Mount Everest
length of a typical car	diameter of the Earth	diameter of the universe
diameter of the Sun	diameter of a quark	diameter of Jupiter

Rearrange the list from smallest to largest.

3 A proton is about a million times larger than an electron. How many orders of magnitude larger is that?

4 Complete the following table:

Particle	Charge (C)	Mass (kg)
Electron		
Proton		
Neutron		
Positron		
Anti-proton		
Anti-neutron		

5 What is the difference between a hadron and a lepton?

6 Name the six leptons.

7 What is a neutrino?

8 Name the six quarks that make up hadrons.

9 What is a boson?

10 What is the difference between a baryon and a meson?

11 Name four bosons and their corresponding force-mediating particles.

12 Use the correct word from the list provided to complete the diagram:

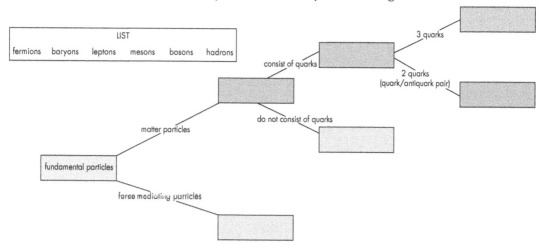

LIST

fermions   baryons   leptons   mesons   bosons   hadrons

consist of quarks

3 quarks

2 quarks (quark/antiquark pair)

do not consist of quarks

matter particles

fundamental particles

force mediating particles

13 What is a quark?

14 List the four fundamental forces in terms of (a) their range (shortest to longest) and (b) their strength (weakest to strongest).

15 What is antimatter and why can it be dangerous?

Example

A sub-atomic particle called a 'charmed Xi' has a quark configuration of 'usc', which means it is made up of an 'up' quark (u), a 'strange' quark (s) and a 'charm' quark (c). Find the charge carried by this particle using the table below (in question 16).

'usc' $Q = + \frac{2}{3} + \frac{2}{3} - \frac{1}{3} = \underline{+1}$

16 The table below shows the symbol and charge for the six different quarks/antiquarks:

QUARKS			ANTIQUARKS		
Quark	Symbol	Charge/amu	Quark	Symbol	Charge/amu
Up	u	$+\frac{2}{3}$	Anti-up	$\bar{u}$	$-\frac{2}{3}$
Down	d	$-\frac{1}{3}$	Anti-down	$\bar{d}$	$+\frac{1}{3}$
Top	t	$+\frac{2}{3}$	Anti-top	$\bar{t}$	$-\frac{2}{3}$
Bottom	b	$-\frac{1}{3}$	Anti-bottom	$\bar{b}$	$+\frac{1}{3}$
Strange	s	$-\frac{1}{3}$	Anti-strange	$\bar{s}$	$+\frac{1}{3}$
Charm	c	$+\frac{2}{3}$	Anti-charm	$\bar{c}$	$-\frac{2}{3}$

Use the table to determine the charge of each of the following particles from its quark constituents:

a Lambda (Λ) uds

b Sigma (Σ) ddb

c Omega (Ω) ccb

d anti-Sigma ($\bar{\Sigma}$) $\bar{u}\,\bar{u}\,\bar{c}$

e anti-Sigma ($\bar{\Sigma}$) $\bar{u}\,\bar{u}\,\bar{b}$

f anti-Lambda ($\bar{\Lambda}$) $\bar{u}\,\bar{d}\,\bar{b}$

g Omega (Ω) sbb

h Sigma (Σ) udc

10 Forces on charged particles

Exercise 10A Force fields

$$m_e = 9{\cdot}11 \times 10^{-31}\ \text{kg} \quad q_e = (-)1{\cdot}6 \times 10^{-19}\ \text{C} \quad m_p = 1{\cdot}673 \times 10^{-27}\ \text{kg} \quad q_p = 1{\cdot}6 \times 10^{-19}\ \text{C}$$

1 The term 'force field' appears in various branches of physics – gravitational fields, electric fields and magnetic fields. What exactly is a 'field' when applied to physics?

2 What are field lines?

3 Re-draw the following combinations including the electric field pattern showing field lines around each particle.

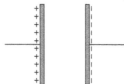

single positive charge   single negative charge   two like charges   two unlike charges

(i)   (ii)   (iii)   (iv)

4 Draw the field lines between the two parallel metal plates below.

Exercise 10B Movement in electric fields

1 Define the 'volt'.

2 Part of a passage in a book on electricity refers to a 'potential difference of 6 V'. What does this mean?

3 How many electrons are there in 1 coulomb of charge?

4 Complete the table below:

Charge	Current	Time
250 mC	50 mA	
	10 μA	20 s
75 C		5 minutes
	2 μA	10 hours
45 C		50 minutes
40 μC	0·5 μA	

5 Complete the table below:

Work	Charge	Voltage
600 mJ	50 mC	
	25 μC	15 V
1 mJ	0·1 C	
100 J		50 V
	0·5 μC	80 kV
45 μJ		3 mV

6 A p.d. of 25 kV is used to accelerate protons from rest to a very high speed.

   a Find the maximum kinetic energy gained by the protons.

   b Find the maximum speed of the protons.

   The accelerating voltage is increased to 35 kV.

   c How does this affect the maximum speed of the protons?

7 A charged particle that has a mass of 4.00×10^{-4} kg reaches a speed of 1000 ms^{-1} when an accelerating potential of 20.0 kV is applied. Calculate the charge of the particle.

8 A charged particle has a charge of 100 mC, a mass of 2.50×10^{-5} kg and is accelerated from rest to 1.25×10^4 ms^{-1}. Calculate the accelerating potential required to gain this speed.

9 In the diagram below, plate A has a potential of +200 V and plate B has a potential of +500 V.

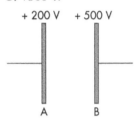

+ 200 V   + 500 V

A   B

How much energy would be required to transfer 3 coulombs of charge between plates A and B?

10 Two parallel conducting plates X and Y are 2.50×10^{-3} m apart in a vacuum. A p.d. of 2.00 kV is applied across the plates.

   a Calculate the kinetic energy gained by an electron in moving from X to Y as shown.

   b Calculate the speed of the electron when it reaches plate Y.

   c Calculate the magnitude of the force exerted by the electric field on the electron as it moves from X to Y.

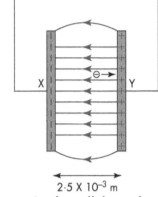

2.5 X 10^{-3} m

11 A particle accelerator is used to accelerate protons between a pair of parallel metal plates.

   a How much kinetic energy is gained by the proton as it moves from plate X to plate Y?

   b The proton has exactly 2.60×10^{-15} J of energy as it passes through a hole in plate X. Calculate the velocity of the proton when it reaches plate Y.

   c The plates are 1.5 m apart. Calculate the force exerted on the proton by the particle accelerator as it moves from plate X to plate Y.

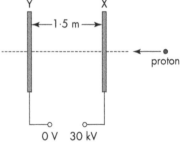

12 A positively charged particle has a charge of 3.20×10^{-19} C. It is accelerated from rest to a speed of 6.91×10^5 ms^{-1} by a potential difference of 5.00 kV.

   a How many protons make up the charged particle?

   b Find the mass of the particle.

   c Identify the particle if the mass of a neutron is approximately 1.677×10^{-27} kg.

   Hint   Using the mass and the charge, work out how many neutrons there are.

13 A crude particle accelerator is manufactured by a university student to accelerate ions between a metal grid and a metal plate.

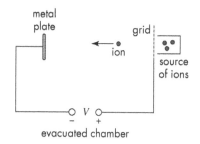

During testing, a proton is accelerated by a potential difference, V, causing the proton to acquire a speed of $2 \cdot 00 \times 10^6 \, \mathrm{ms^{-1}}$ upon reaching the metal plate. During a second test the same potential difference is used to accelerate an alpha particle that consists of 2 protons and 2 neutrons (total mass $6 \cdot 64 \times 10^{-27}$ kg).

Calculate the speed of the alpha particle when it reaches the metal plate.

> **Hint** Remember the charge on an alpha particle is $+2q_p$.

Exercise 10C Applications of electric fields

1 Charges pass between two parallel plates. Copy the diagrams and insert the electric field lines and the path taken by the charge in each case.

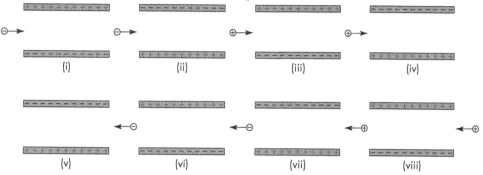

2 An oscilloscope consists of an electron 'gun', two sets of deflection plates and a phosphor screen encapsulated within a vacuum tube. The electron 'gun' is used to accelerate electrons from rest towards the deflection plates and the phosphor screen.

The electron 'gun' consists of a cathode, a grid and two anodes as shown.

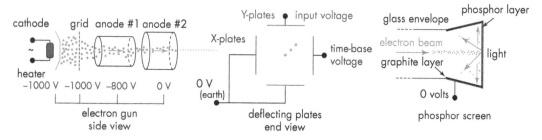

The cathode has a p.d. of 1000 V relative to the second anode.

a Calculate the speed of the electrons on reaching:

   i the grid;   **ii** the first anode; and   **iii** the second anode.

> **Hint** The speed will increase as the result of an increase in the p.d. between each point.

b The distance between the cathode and the anode is 100 mm. Calculate the acceleration of the electrons over this distance.

c What is the purpose of (i) the X-plates and (ii) the Y-plates?

d How could the deflection to the top right (as shown in the X- and Y- plates part of the image) be achieved?

e Why is a vacuum tube required?

Exercise 10D Moving charges in a magnetic field

 Copy the images below and draw in the magnetic field lines around the magnets in each case.

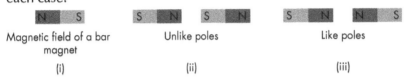

Magnetic field of a bar magnet

(i)

Unlike poles

(ii)

Like poles

(iii)

Example

An electron is about to enter into a magnetic field as shown below. Find the direction that the electron moves within the field.

magnetic field into page

- First, as the particle is an **electron** we need to use the **RIGHT-HAND RULE.** This involves orientating your right hand according to the image shown at the bottom with your thumb and first two fingers at 90° to each other.

- Next, point your first finger in the direction of the electric field, i.e. into the page in this case.

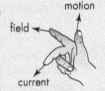

Right-hand rule

- Next, orientate your right hand (keeping the field direction unchanged) until your second finger points in the direction of the charge motion before it enters the field, i.e. left to right here.

- Finally, the direction of your thumb indicates the direction in which the charge will move in the field, which for this example would be towards the bottom of the page.

 For each of the images below, determine the direction of the charge as it passes through the magnetic field.

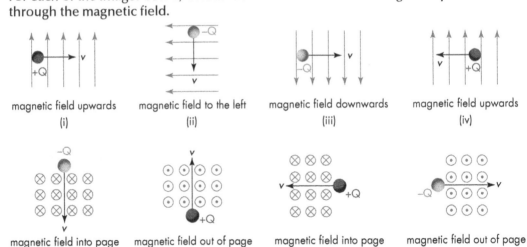

| magnetic field upwards (i) | magnetic field to the left (ii) | magnetic field downwards (iii) | magnetic field upwards (iv) |

| magnetic field into page (v) | magnetic field out of page (vi) | magnetic field into page (vii) | magnetic field out of page (viii) |

3 Work out the direction the current-carrying wire in each case below will move.

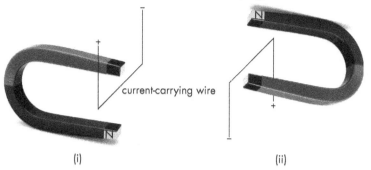

(i)   (ii)

> **Hint**   Remember: right-hand rule for –ve charges and left-hand rule for +ve charges.

Exercise 10E Particle accelerators

1 Name three types of particle accelerator commonly used by particle physicists.

2 In two of these particle accelerators both electric and magnetic fields are used.
What is the purpose of (a) the electric field and (b) the magnetic field?

3 A linear accelerator consists of cylinders (called drift tubes) of increasing length as shown:

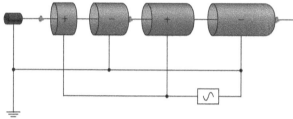

The cyliders are connected to an alternating supply as shown.

a Why are the cylinders increasing in length?

b Why does the accelerator need an alternating supply?

c Whereabouts are the particles (i) accelerating and (ii) moving with constant velocity?

4 Using the diagram, explain how a cyclotron moves a proton from the centre to the target.

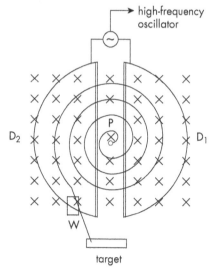

11 Nuclear reactions

Exercise 11A Model of the atom

1 What is meant by (a) the atomic number and (b) the mass number?

2 By how many orders of magnitude is the atom larger than an atomic nucleus?

3 Write down the mass number, charge and symbol used for the three particles that make up an atom.

4 What is the collective name used for the constituents of the nucleus?

5 What determines the position of an element in the periodic table?

6 What are isotopes?

7 What is the relationship between the number of neutrons, the mass number and the atomic number?

8 For each of the elements below, state the number of neutrons present in the nucleus:

a $^{98}_{43}\text{Tc}$

b $^{55}_{25}\text{Mn}$

c $^{190}_{76}\text{Os}$

9 a Identify the following three elements (periodic table required).

   i $^{56}_{26}$?   ii $^{238}_{92}$?   iii $^{96}_{42}$?

   b How many protons and neutrons does each element have?

10 The nucleus contains protons, which will normally repel each other, so why doesn't the nucleus fall apart?

11 What is meant by the term 'ionisation'?

12 During radioactive decay, where does a β-particle emanate from? Explain.

Exercise 11B Binding energy

1 Identify the missing values 'x', 'y' and 'z' in each of the following nuclear reactions **without** using a periodic table:

a $^{x}_{0}\text{n} + ^{235}_{92}\text{U} \rightarrow ^{139}_{y}\text{La} + ^{z}_{42}\text{Mo} + 2\,^{1}_{0}\text{n}$

b $^{1}_{0}\text{n} + ^{x}_{92}\text{U} \rightarrow ^{137}_{54}\text{Xe} + ^{94}_{y}\text{Sr} + 3\,^{1}_{0}\text{n}$

c $^{1}_{0}\text{n} + ^{235}_{92}\text{U} \rightarrow ^{140}_{x}\text{Ce} + ^{94}_{40}\text{Zr} + y\,^{1}_{0}\text{n}$

2 What is meant by the term 'binding energy'?

Example

The decay of uranium-238 into thorium-234 is shown here: $^{238}_{92}U \rightarrow {}^{234}_{90}Th + {}^{4}_{2}He$

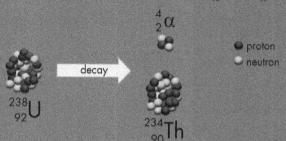

$^{4}_{2}\alpha$

● proton
◔ neutron

$^{238}_{92}U$

decay

$^{234}_{90}Th$

Mass of U-238 = $3{\cdot}983 \times 10^{-25}$ kg

Mass of Th-234 = $3{\cdot}887 \times 10^{-25}$ kg

Mass of He-4 = $6{\cdot}642 \times 10^{-27}$ kg

Calculate the energy released during this decay.

- First, add together all the masses to the left of the arrow – in this case there is only one – $3{\cdot}983 \times 10^{-25}$ kg

- Next, add together all the masses to the right of the arrow, i.e. $3{\cdot}887 \times 10^{-25} + 6{\cdot}642 \times 10^{-27} = 3{\cdot}95342 \times 10^{-25}$ kg – **Important:** do **NOT** round at this stage!

- Next, subtract the total mass on the right of the arrow from the total mass on the left:

  Mass difference = $3{\cdot}983 \times 10^{-25} - 3{\cdot}95342 \times 10^{-25} = 2{\cdot}958 \times 10^{-27}$ kg

- Finally, use $E = mc^2$ where m is the mass difference calculated above and c is the speed of light.

  Energy released, $E = 2{\cdot}958 \times 10^{-27} \times (3{\cdot}00 \times 10^{8})^2 = \underline{2{\cdot}66 \times 10^{-10}}$ J

3 For each of the following nuclear reactions, find (i) the mass difference before and after a nucleus splits and (ii) the binding energy in each case.

a $^{1}_{0}n + {}^{235}_{92}U \rightarrow {}^{137}_{55}Cs + {}^{95}_{37}Rb + 4\,{}^{1}_{0}n$

b $^{240}_{94}Pu \rightarrow {}^{236}_{92}U + {}^{4}_{2}He$

c $^{1}_{0}n + {}^{235}_{92}U \rightarrow {}^{140}_{58}Ce + {}^{94}_{40}Zr + 2\,{}^{1}_{0}n$

d $^{3}_{1}H + {}^{2}_{1}H \rightarrow {}^{4}_{2}He + {}^{1}_{0}n$

e Where would you find the reaction in part (d) above occurring in abundance?

Particle	Mass ($\times\,10^{-27}$ kg)
$^{1}_{0}n$	1·675
$^{1}_{1}p$	1·673
$^{235}_{92}U$	390·173
$^{137}_{55}Cs$	227·292
$^{95}_{37}Rb$	157·562
$^{240}_{94}Pu$	398·626
$^{236}_{92}U$	391·970
$^{4}_{2}He$	6·645
$^{140}_{58}Ce$	232·242
$^{94}_{40}Zr$	155·884
$^{3}_{1}H$	5·005
$^{2}_{1}H$	3·342
$^{1}_{1}H$	1·674
$^{4}_{2}He$	6·645
$^{3}_{2}He$	5·008

4 Helium-3 (^3_2He) can be used to produce helium-4 (^4_2He) in the following reaction:

$$^3_2\text{He} + {}^2_1\text{H} \rightarrow {}^4_2\text{He} + {}^1_1\text{p}$$

Calculate the energy released in this reaction using information in the table on the previous page.

5 A reactor requires 25 GW of power from the reaction in question 3 part (a).

How many nuclear reactions are required each second?

6 A reactor requires 40 GW of power from the reaction in question 3 part (c).

How many nuclear reactions are required each minute?

7 When neutron induced fission of U-235 takes place, a variety of different products can be produced. What is it that determines the products in this type of reaction?

8 Explain briefly why energy is released in the following reaction:

$$^1_0\text{n} + {}^{235}_{92}\text{U} \rightarrow {}^{139}_{56}\text{Ba} + {}^{94}_{36}\text{Kr} + 3\,{}^1_0\text{n} + energy$$

Exercise 11C Radioactive decay

1 What is:

a an α-particle;

b a β-particle; and

c a γ-ray?

2 Three ionising radiations are fired at three materials.

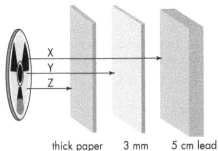

thick paper   3 mm aluminium   5 cm lead

Which radiation is α, β and γ?

3 Complete the missing information in the following table:

Name	Symbol	Nature	Charge (amu)	Mass (nucleon units)	Speed	Stopped by
Alpha	$^4_2\alpha$			4		
		fast moving electron	−1		90% of light	
Gamma		e-m radiation		0		many cm of lead

4 What is the symbol for:

a an α-particle; and

b a β-particle?

(Include mass and atomic numbers.)

5 What happens to:
   a the atomic number; and
   b the mass number of a radioisotope when an α-particle is released?

6 What happens to:
   a the atomic number; and
   b the mass number of a radioisotope when a β-particle is released?

7 What happens to:
   a the atomic number; and
   b the mass number of a radioisotope when a γ-ray photon is released?

Example

Shown below is part of the thorium-232 decay chain. Find the particles liberated at each stage.

$$^{228}_{89}\text{Ac} \rightarrow {}^{228}_{90}\text{Th} \rightarrow {}^{224}_{88}\text{Ra}$$

- First, use simple arithmetic treating the arrow like an equals sign and find the difference between the mass numbers and the difference between the atomic numbers to find the total mass and atomic number of the particle(s) released at each stage.

$$^{228}_{89}\text{Ac} \rightarrow {}^{228}_{90}\text{Th} + {}^{\text{mass number}}_{\text{atomic number}}\text{particle} \quad \text{mass number} = 0, \text{ atomic number } -1$$

hence the particle released is a <u>β-particle</u>.

- Repeat the process with the next part:

$$^{228}_{90}\text{Th} \rightarrow {}^{224}_{88}\text{Ra} + {}^{\text{mass number}}_{\text{atomic number}}\text{particle} \quad \text{mass number} = 4, \text{ atomic number } 2$$

hence the particle released is an <u>α-particle</u>.

8 Part of a radioactive decay chain is shown below:

$$^{218}_{84}\text{Po} \rightarrow {}^{214}_{82}\text{Pb} \rightarrow {}^{214}_{83}\text{Bi} \rightarrow {}^{214}_{84}\text{Po} \rightarrow {}^{210}_{82}\text{Pb}$$

Identify the radiation released at each stage.

9 How easy is it to predict precisely when a radioactive nuclide will undergo decay?

10 Part of a decay series starting with uranium, $^{238}_{92}\text{U}$, involves the emission in turn of an alpha, beta, beta, alpha, alpha, alpha, alpha.
Identify the final product formed at the end of this part of the decay.
(You will need a periodic table for this.)

11 Part of a decay sequence is shown in the image here.

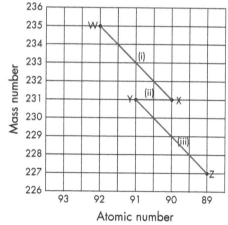

a Identify the three particles released in this part of the decay sequence.

b Identify each of the elements W, X, Y and Z. (You will require a periodic table for this.)

c Explain why the image shown does not provide a complete picture of the decay process.

12 A condensed part of a chain reaction is shown below:

$$^{230}_{90}\text{Th} \rightarrow \ ^{218}_{84}\text{Po} + 3 \text{ particles}$$

Identify the three particles using only α and β-particles.

13 A condensed part of a chain reaction is shown below:

$$^{210}_{82}\text{Pb} \rightarrow \ ^{206}_{82}\text{Pb} + 3 \text{ particles}$$

Identify the three particles using only α and β-particles.

14 A condensed part of a chain reaction is shown below:

$$^{222}_{86}\text{Rn} \rightarrow \ ^{214}_{83}\text{Bi} + 3 \text{ particles}$$

Identify the three particles using only α and β-particles.

Exercise 11D Nuclear fission and nuclear fusion

1 Describe the difference between nuclear fission and nuclear fusion.

2 Give three reasons why nuclear fusion is a preferred means of generating energy to nuclear fission.

3 What is meant by a chain reaction?

4 Why does unmined uranium-235 not undergo fission, creating nuclear explosions, in nature?

5 How is $^{4}_{2}\text{He}$ created in the Sun?

6 Enriched uranium contains 2·3% U-235 and is used in nuclear weapons. Each fission releases $3{\cdot}36 \times 10^{-11}$ J of energy. Calculate the total energy released when all the U-235 in 10 kg of natural uranium-235 has undergone fission.

7 An image of a potential nuclear fusion reactor is shown here.

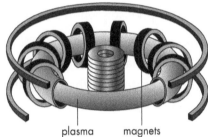

plasma   magnets

   a What exactly is nuclear plasma?

   b Why are electromagnets used to surround the plasma?

   c Why is the plasma confinement vessel shaped like that shown above?

   d Why has mankind so far failed to produce a sustainable fusion reactor?

12 The photoelectric effect

Exercise 12A Photon energy

$$h = 6 \cdot 63 \times 10^{-34} \text{ J s} \qquad m_e = 9 \cdot 11 \times 10^{-31} \text{ kg}$$

1 Describe what is meant by a 'photon'.

2 Write down an expression for the energy of a photon in terms of its wavelength. Explain the meaning of each term used and the quantities that it is measured in.

3 Complete the missing values in the following table:

Frequency ($\times 10^{14}$ Hz)	Wavelength (nm)	Energy ($\times 10^{-19}$ J)
	250	7·96
9·38		6·22
6·38	470	
5·45		3·61
4·92	610	
	750	2·65

4 Which member of the electromagnetic spectrum has photons of:

a the shortest wavelength;

b the lowest frequency; and

c the greatest energy?

5 A red laser emits radiation with a wavelength of 630 nm and has a power rating of 2·50 W.

a How much energy is delivered by the laser in 8·0 s?

b Calculate the number of photons released by this laser to provide the 8·0 s pulse of energy.

6 A 2·2 W krypton gas laser is used to repair detached retinas. It emits radiation with a frequency of $6 \cdot 38 \times 10^{14}$ Hz and the laser produces pulses of radiation that last for 49 ms.

a What is the wavelength of this laser?

b What is the colour of the laser beam?

c How much energy in joules is delivered by the laser in 49 ms?

d Calculate the number of photons released by this laser to provide the 49 ms pulse of energy.

7 The photon energies of three different sources are shown below:

• Radiation 1: $3 \cdot 14 \times 10^{-19}$ J

• Radiation 2: $5 \cdot 23 \times 10^{-19}$ J

• Radiation 3: $6 \cdot 47 \times 10^{-19}$ J

A student makes the following four statements about these radiations:

a Radiation 2 has a higher frequency than radiation 3.

b Radiation 1 has a longer wavelength than radiation 2.

c Radiation 3 has a shorter wavelength than radiation 2.

d Radiation 2 has a lower frequency than radiation 1.

Which of these statements is/are correct?

Exercise 12B The photoelectric effect

1 What is meant by the 'photoelectric effect'?

2 What is meant by 'photoemission'?

3 For each of the following situations decide whether or not the gold-leaf electroscope will show that photoemission has occurred.

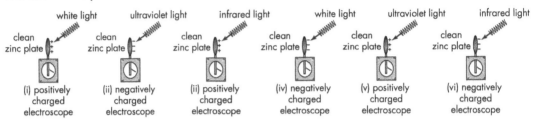

white light	ultraviolet light	infrared light	white light	ultraviolet light	infrared light
clean zinc plate	clean zinc plate	clean zinc plate	clean zinc plate	clean zinc plate	clean zinc plate
(i) positively charged electroscope	(ii) negatively charged electroscope	(ii) positively charged electroscope	(iv) negatively charged electroscope	(v) positively charged electroscope	(vi) negatively charged electroscope

4 A clean zinc plate photoemits with ultraviolet light but a dirty zinc plate does not. Explain both of these situations.

5 Radiation is shone onto a negatively charged gold-leaf electroscope and no photoemission takes place. What will happen if the radiation used has its intensity increased?

6 Faint green light and intense red light are both shone onto a negatively charged metal plate sitting on top of a gold-leaf electroscope. Photoemission occurs with both light sources.

  a Explain why the faint green light photons release photoelectrons with greater kinetic energies.

  b Explain which source will produce more emitted photoelectrons.

7 Explain what is meant by each of the following terms:

  a Threshold frequency

  b Work function

  c Threshold wavelength

Example

The work function of mercury is $7 \cdot 21 \times 10^{-19}$ J. An ultraviolet lamp with a wavelength of 249 nm is shone onto the surface of mercury. Determine whether any photoelectrons are released and if so, calculate the average speed of the escaping electrons.

- First, calculate the frequency of the ultraviolet light used:

$$f = \frac{c}{\lambda} = \frac{3 \cdot 00 \times 10^8}{249 \times 10^{-9}} = 1 \cdot 20 \times 10^{15} \text{ Hz}$$

- Next, use the appropriate equation to determine whether there is any kinetic energy available to allow the photoelectrons to escape:

$$E_k = hf - hf_o = (6 \cdot 63 \times 10^{-34} \times 1 \cdot 20 \times 10^{15}) - 7 \cdot 21 \times 10^{-19} = \underline{7 \cdot 46 \times 10^{-20} \text{ J, which is greater than the work function therefore photoelectrons are emitted.}}$$

- Finally, equate $E_k = \frac{1}{2} mv^2$ to the E_k of the escaping photoelectrons to find v:

$$7 \cdot 46 \times 10^{-20} = \frac{1}{2} \times 9 \cdot 11 \times 10^{-31} \times v^2$$

$$v = \underline{4 \cdot 05 \text{ ms}^{-1}}$$

8 A clean tin plate has a work function of $7 \cdot 1 \times 10^{-19}$ J. Radiation of frequency $1 \cdot 4 \times 10^{15}$ Hz is shone onto the plate and photoemission occurs.

   a Calculate the energy of the incident radiation.

   b Calculate the energy of the escaping photoelectrons.

   c Calculate the maximum speed of the escaping photoelectrons.

9 Calcium metal requires photons with an energy of $4 \cdot 65 \times 10^{-19}$ J to eject an electron from the metal's surface.

   a Calculate the threshold frequency of radiation required to emit a photoelectron from the metal.

   b Explain whether or not photoemission would take place using radiation of:

      i frequency $6 \cdot 20 \times 10^{14}$ Hz; and

      ii wavelength 410 nm.

10 Photoemission occurs when light is shone onto a clean zinc surface. The work function of the zinc is $5 \cdot 81 \times 10^{-19}$ J. The kinetic energy of the escaping photoelectrons is $2 \cdot 146 \times 10^{-19}$ J. Find:

   a the frequency of the incident photons;

   b the wavelength of the incident photons; and

   c the maximum speed of the escaping photoelectrons.

11 A metal has a work function of $7 \cdot 43 \times 10^{-19}$ J.

   a What is the metal's threshold frequency?

   b What is the metal's threshold wavelength?

12 Calcium has a work function of $4 \cdot 6 \times 10^{-19}$ J. Light shone onto a clean calcium disc releases photoelectrons with a maximum kinetic energy of $1 \cdot 2 \times 10^{-19}$ J. Find the wavelength of the light used.

13 A metal is used to produce photoelectrons when light of wavelength 350 nm is shone onto it. The maximum kinetic energy of the escaping photoelectrons is $2 \cdot 01 \times 10^{-19}$ J. Which of the following metals was used?

Metal	Work function (J)
sodium	$3 \cdot 78 \times 10^{-19}$
potassium	$3 \cdot 67 \times 10^{-19}$
barium	$4 \cdot 04 \times 10^{-19}$

14 A metal is used to produce photoelectrons when light of wavelength 420 nm is shone onto it. The maximum kinetic energy of the escaping photoelectrons is $3 \cdot 70 \times 10^{-19}$ J.

   a What is the threshold wavelength?

   b What is the threshold frequency?

15 A technician carries out an investigation involving the photoelectric effect using the apparatus shown here.

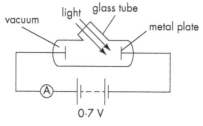

The metal plate on the right (the cathode) is coated with cadmium. Radiation with a frequency of 1.03×10^{15} Hz is shone onto this cathode, causing it to emit photoelectrons.

a Calculate the maximum kinetic energy of the liberated photoelectrons if the threshold frequency of cadmium is 6.520×10^{-19} Hz.

b A photoelectron leaves the cadmium-coated electrode (cathode) with this maximum kinetic energy. Calculate its kinetic energy when it reaches the left-hand electrode (anode) when the p.d. across the electrodes is 0.70 V.

c The polarity of the supply voltage is now reversed. Calculate the minimum voltage required to stop photoelectrons reaching the anode.

16 Photoelectrons with a maximum speed of 5.13×10^5 ms^{-1} are released from a metal after light with a frequency of 8.75×10^{14} Hz is shone onto the metal's surface. Calculate the work function of the metal.

17 The apparatus shown here is used to demonstrate the photoelectric effect.

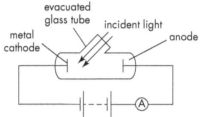

A photoelectric current is measured using the ammeter.

a Draw a graph showing how the photoelectric current varies with frequency of incident light.

b Draw a graph showing how the photoelectric current varies with intensity of incident light.

c Draw a graph showing how the photoelectric current varies with applied voltage.

18 Ultraviolet radiation is shone onto a clean zinc plate causing photoemission to occur. The maximum kinetic energy of the photoelectrons and the rate at which they are released is measured. The irradiance of the ultraviolet radiation is now increased.

a What impact will this have on the maximum kinetic energy of the photoelectrons?

b How will the rate at which the photoelectrons are released be affected?

19 A student makes several statements about the photoelectric effect:

i If the frequency of incident radiation is less than the threshold frequency, decreasing the intensity will decrease the maximum kinetic energy of the escaping photoelectrons.

ii If the frequency of the incident radiation is more than the threshold frequency, increasing the intensity will increase the number of photoelectrons released.

iii Photoelectric emission from a metal only occurs when the frequency of the incident radiation is less than the threshold frequency.

iv Photoelectric emission takes place when the metal's temperature is increased.

v Increasing the intensity of a light source has no effect upon whether the metal emits photoelectrons or not.

vi Increasing the intensity (assuming photoemission takes place) increases the kinetic energy of the escaping photoelectrons.

vii Increasing the intensity will increase the number of photoelectrons liberated, provided photoemission occurs.

viii Increasing the frequency of the incident photons (assuming photoemission takes place) increases the kinetic energy of the liberated photoelectrons.

    a Which of these statements is true?

    b By making slight changes, correct the false statements so that they now become true statements.

20 What is meant by the 'stopping potential' when applied to the photoelectric effect?

21 Calculate the required stopping potential, V_s, for photoelectrons that are emitted when incident light of frequency $1{\cdot}4 \times 10^{15}$ Hz strikes the surface of a metal that has a work function of $4{\cdot}00 \times 10^{-19}$ J.

> **Hint**   Remember that $E = qV$.

22 A student is carrying out an experiment on the photoelectric effect and makes several measurements of the maximum kinetic energy of escaping electrons as the frequency of incident radiation is varied. The graph shown here is produced.

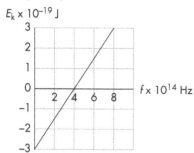

> **Hint**   Think of the mathematical equation for a straight line graph and equate it to the Planck–Einstein equation.

Use the graph to:

a find the threshold frequency;

b find the value of Planck's constant from the experiment; and

c find the work function of the metal used.

13 Inverse square law

Exercise 13A Inverse square law

> Area of a sphere $= 4\pi r^2$

1 What is the difference between the 'brightness' and 'irradiance' of a lamp?

Example

A small point light source has an irradiance of $4.0\,\mathrm{W\,m^{-2}}$ when situated $0.50\,\mathrm{m}$ above a floor. What will be the new irradiance if the source is now moved to $2.5\,\mathrm{m}$ above the floor?

- First, use an appropriate equation:

$I = \dfrac{k}{d^2}$, which is better written as $I_1 d_1{}^2 = I_2 d_2{}^2$ where '1' represents before any change and '2' represents after the change has occurred.

- Finally, insert the values into the equation and solve:

$I_1 d_1{}^2 = I_2 d_2{}^2$

$4.0 \times 0.50^2 = I_2 \times 2.5^2$

$I_2 = \underline{0.16\,\mathrm{W\,m^{-2}}}$

2 Find the missing values in the following table:

Irradiance, I_1 (W m⁻²)	Distance, d_1 (m)	Irradiance, I_2 (W m⁻²)	Distance, d_2 (m)
2.0	4.0	8.0	
2.5	0.25		2.5
4.0		64.0	2.5
	5.0	0.025	0.50
0.025	0.50		0.05
100.0	2.0	4.0	
0.50		2.0	16.0
0.225	16.0		3.0

3 A small point light source has an irradiance 'I' when situated $2.5\,\mathrm{m}$ above a floor. What will be the new irradiance if the source is now moved to $5.0\,\mathrm{m}$ above the floor?

4 A light meter placed $10\,\mathrm{cm}$ from a point light source measures an irradiance of $4\,\mathrm{mW\,m^{-2}}$. The light meter is moved and the irradiance measured is now $64\,\mathrm{mW\,m^{-2}}$. How far (in cm) from the light meter is the point light source now?

5 A point light source and a light meter together with a metre rule are used to investigate how the irradiance varies with distance in a darkened room. The results are shown below:

Irradiance (units)	1250	313	139	78	50
Distance (m)	0·20	0·40	0·60	0·80	1·00

a Using graph paper, plot a graph of irradiance against 1/distance².

b Use your graph to determine what the irradiance would be at a distance of 0·25 m from the point light source.

c What is the value of the constant of proportionality?

d On your graph draw a dashed line to show what would be the result of repeating the experiment in the same room with the lights on.

6 A point light source uniformly emits $1·4 \times 10^{12}$ photons per second in all directions. Assuming the cross sectional area of a student is $1·0 \, m^2$, estimate the number of photons arriving per second when the student is standing 3 m away from the light source.

> Hint Calculate the area of a sphere and work out what fraction of this is covered by the person.

7 The solar panels of a space probe are $2·2 \times 10^6$ km from the Sun. The space probe requires solar panels with a surface area of $10 \, m^2$ in order to produce enough power to operate the probe. The space probe is to be repositioned at a distance of $4·4 \times 10^6$ km from the Sun. What area of solar panels would be required for the probe to operate as before?

> Hint First work out what happens to the irradiance.

8 The irradiance of the Sun at its surface is $6·33 \times 10^7 \, W\,m^{-2}$. The mean diameter of the Sun is $1·3914 \times 10^6$ km. What will be the irradiance at the Earth's surface if the mean distance from the centre of the Sun is $149·6 \times 10^6$ km?

> Hint Work out the initial distance of the Sun's surface first.

14 Interference and diffraction

Exercise 14A Diffraction

$$v_{sound} \text{ (in air)} = 340\,\text{m}\,\text{s}^{-1}$$

1. Define what is meant by the term 'diffraction'.

2. Copy the following diagrams and draw the wavefronts after they pass through the barrier.

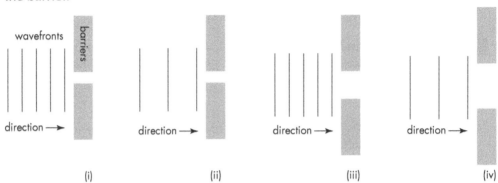

(i)   (ii)   (iii)   (iv)

Exercise 14B Interference

1. Explain what is meant by the term 'constructive interference'.

2. Explain what is meant by the term 'destructive interference'.

3. Explain what is meant by 'constant phase difference' when applied to waves.

4. Explain what is meant by the term 'coherent' when applied to waves.

5. A slinky is used to produce identical pulses as shown.

           (i)   (ii)

   a Draw diagrams showing what would happen in each case when the two pulses meet.

   b Which of the two diagrams will result in destructive interference when the pulses meet?

6. Copy the sketch shown here and draw on the same diagram a wave that is 180° out of phase with the sketch shown.

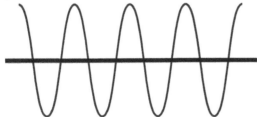

7 A laser is used to produce a narrow beam of monochromatic green light, which is passed through a narrow double slit onto a screen.

a Explain how the fringe pattern is produced.

b Describe what would happen to this fringe pattern if the monochromatic green laser was replaced with one which emitted monochromatic blue light. Explain.

c Describe what would happen to this fringe pattern if the monochromatic green laser was replaced with one which emitted monochromatic red light. Explain.

Example

A signal generator and two identical loudspeakers are used to produce a note with a frequency of 2·125 kHz. An audible interference pattern is then produced. What is the nature of the interference produced at point X and what is its order number?

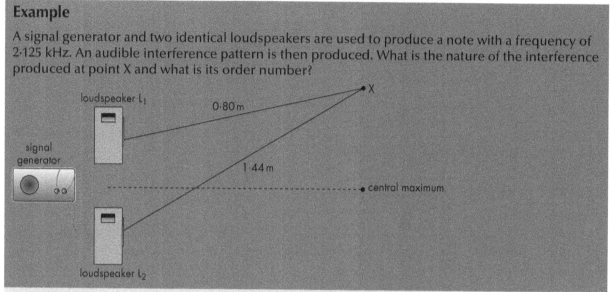

- First, use an appropriate equation to find the wavelength of the sound used:

$v = f\lambda$

$340 = 2125 \times \lambda$

$\lambda = 0.16$ m

- Next, use an appropriate equation, path difference $= m\lambda$, to establish if m is a whole number. If so, then it will be a point of constructive interference. If not, then it will be a point of destructive interference.

- Finally, insert values:

$L_1X - L_2X = m\lambda$

$1.44 - 0.80 = m \times 0.16$

$\underline{m = 4}$ and since it is a whole number it must be a point of <u>constructive interference</u>.

8 A microwave source produces an interference pattern as shown here. The waves meet at the second order maxima ($m = 2$) as shown.

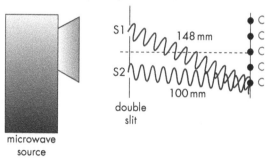

Calculate the wavelength of the microwave source.

9 The same microwave source in question 8 produces another interference pattern as shown here. This time the waves meet at a zero order minimum.

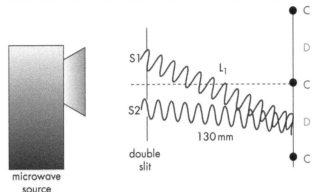

Calculate the distance L_1 from slit S_1 to the zero order minimum.

10 A microwave source emits a frequency of 1.07×10^{10} Hz and produces an interference pattern when passed through a double slit. The path lengths of the two waves from the slits to a detector are 266 mm and 168 mm. What is the nature of the interference pattern produced at the detector?

11 Two coherent sound sources with a wavelength of 170 mm are used to produce an interference effect. What would be the path difference where the third order minimum from the central maximum is detected?

12 A 28 mm microwave source produces an interference pattern as shown. What is the nature of the interference pattern produced and what is its order number?

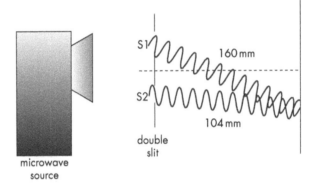

13 Two identical loudspeakers placed 1 m apart are connected to the same signal generator as shown.

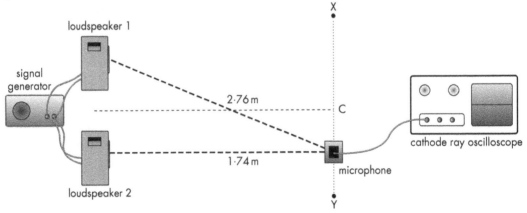

The signal generator is switched on and the microphone moved between X and Y until it is sitting at the second order minimum from point 'C', which is the central maximum.

a Calculate the wavelength of the sound waves produced by the signal generator.

b Calculate the frequency of the sound waves produced by the signal generator.

c What would happen to the spacing between points of constructive interference along XY if the loudspeakers are placed closer together?

d The microphone is still sitting at the second order minimum when loudspeaker 1 is disconnected. Describe and explain what happens to the signal observed on the oscilloscope.

14 Two identical loudspeakers are connected to the same signal generator and both produce a sound that is in phase with one other, producing an interference pattern.

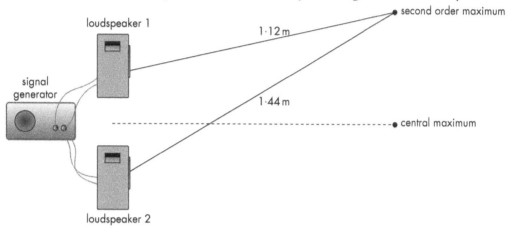

What is the frequency of the sound emitted by the two loudspeakers?

15 Microwaves pass through two barriers creating an interference pattern. One path taken by the waves is shown. The waves have a wavelength of 28 mm. Explain whether point X is a maximum or a minimum.

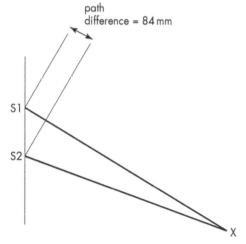

16 A vibrating dipper in a ripple tank produces circular wavefronts with a wavelength of 10 mm at point 'A' as shown.

Part of the wavefronts reflect off a metal barrier at point 'B' and interfere constructively with incoming wavefronts at point 'C'. If AC is 0·18 m, suggest three possible values for the distance ABC.

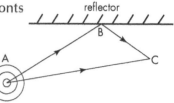

17 Another experiment using a ripple tank, this time with two dippers 'X' and 'Y' moving in and out of the water in phase with each other, produces an interference minimum at point 'Z'.

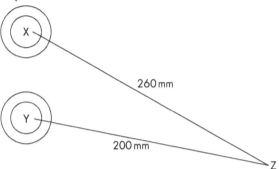

The distance XZ is 260 mm and the distance YZ is 200 mm. Which of the following values could be the wavelength of the waves?

20, 30, 40, 50, 60 mm

18 Two dippers are used in a ripple tank to produce circular overlapping wavefronts. Maxima and minima are produced a short distance away from the two sources as shown.

How will the distribution of maxima and minima change if:

a the two sources are moved further apart without changing the frequency of the sources;

- maximum
- minimum
- central maximum
- minimum
- maximum

b the frequency of the circular wavefronts is decreased and the speed remains constant; and

c the speed of the wavefronts is increased and frequency remains constant?

19 A coherent source of sound waves emitted by two identical loudspeakers form an interference pattern as shown.

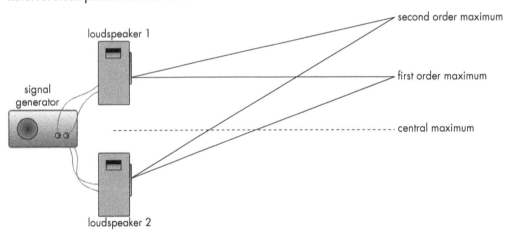

The path length from loudspeaker 1 to the first order maximum is 360 mm. The path length from loudspeaker 2 to the first order maximum is 400 mm.

a Find the frequency of the sound source.

b Find the path difference for the second order maximum.

20 A diffraction grating used in a Young's slits experiment has 300 lines per mm. What is the distance between adjacent lines?

21 The distance between adjacent slits on a diffraction grating is 1.25×10^{-5} m. How many lines per mm does it have?

22 A 300 lines per mm diffraction grating is used with a 670 nm laser to produce an interference pattern on a wall.

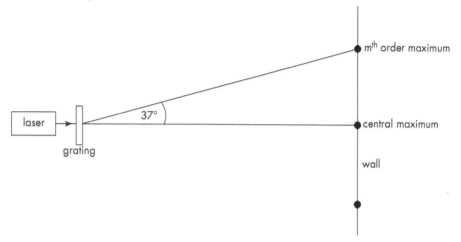

If the angle between the central maximum and the mth order maximum is 37°, find the value of 'm'.

23 A 600 lines per mm diffraction grating is being used with a laser to produce an interference pattern. If the angle between the central maximum and the second order maximum is 51°, calculate the wavelength of the laser used.

24 A laser that has a wavelength of 610 nm is used with a diffraction grating to produce an interference pattern. The angle between the central maximum and the sixth order maximum is 17°. Calculate the number of lines per mm on the diffraction grating used.

25 A 556 nm laser is used with a 300 lines per mm grating to produce an interference pattern on a wall. What will be the angle between the central maximum and the third order maximum?

26 Describe fully the interference pattern produced if white light is used with a diffraction grating to produce the interference pattern.

27 A laser is used to produce an interference pattern on a wall using an 80 lines per mm grating as shown.

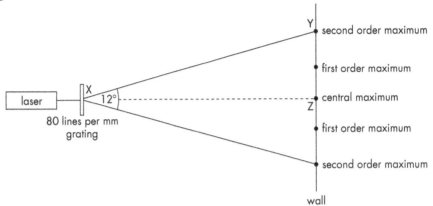

a The distance XY is measured several times and the results are:

1·28 m   1·23 m   1·25 m   1·27 m   1·25 m   1·22 m

Calculate:

i the mean value for the distance XY; and

ii the approximate random uncertainty in this value.

b Distance XZ is $(1·00 \pm 0·01)$ m

Distance YZ is $(0·75 \pm 0·01)$ m

Determine which distance (XY, XZ or YZ) has the largest percentage uncertainty.

c Calculate the wavelength of the light used expressing your answer in the form:

(wavelength ± **absolute** uncertainty)

28 The positions of the maxima fringes in a diffraction pattern depend upon two factors:

i the width of the slit (d); and

ii the wavelength of the waves (λ).

A relationship for the angular position of a maximum is described by the following equation:

$$\sin \theta = \frac{m\lambda}{d} \qquad \text{where } m \text{ is the order}$$

a Calculate the angular position of the second order maximum for a diffraction pattern created when a 500 nm monochromatic light source passes through a single slit of width 0·012 mm.

b Explain why it is better to have as large a distance as possible between the slit and the screen when trying to observe the fringes.

29 A Young's slits experiment is carried out using a laser and a diffraction grating to produce interference fringes on a screen. State the effect on the fringe spacing of:

a increasing the distance between adjacent slits on the diffraction grating;

b using a diffraction grating with a greater number of lines per mm;

c decreasing the distance between the grating and the screen; and

d using a laser with a longer wavelength.

30 A laser that has a wavelength of 670 nm is used with a 300 lines per mm diffraction grating to produce an interference pattern on a screen that is 2·5 m away from the grating. The diffraction grating is then replaced by an 80 lines/mm grating. Find the distance between the two second order maxima formed with each grating.

> **Hint** Find θ for both gratings and then use trigonometry to find the fringe spacing for each.

31 A 28 mm microwave transmitter is used to produce an interference pattern using three metal barriers to form a double slit. The distance from the barriers to the microwave receiver is 500 mm.

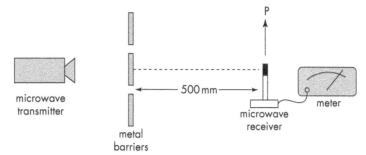

The receiver starts off at the central maximum and moves towards position 'P' which is two maxima further on. A second microwave transmitter with a different wavelength replaces the original transmitter and it detects its second minimum at point 'P'. What is the wavelength of the second microwave source?

> **Hint** Find the path difference to point 'P' – this will be the same for both transmitters.

32 A white light source is used with an 80 lines per mm grating to create an interference pattern on a wall that is 1·25 m away from the grating. The central maximum is white and on each side of the central maximum two coloured spectra can be observed. Given that the two extremes of the visible spectrum are taken to be 400 nm and 700 nm, find the width of the second order spectra formed.

> **Hint** Work out the two angles for the two extremes of the visible spectrum and then use trigonometry.

15 Refraction of light

Exercise 15A Snell's law and refractive index

Colour of light	Wavelength (nm)
Orange	590–640
Yellow	550–580
Green	490–530
Blue	460–480

1 Define the terms 'refraction' and 'normal' when applied to geometrical optics.

2 Define 'refractive index'.

3 Identify the angles labelled (i) → (iv) in the diagram shown here.

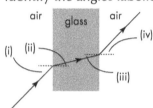

Example

Calculate the angle θ_2 in the image shown here if the refractive index of the block is 1·51.

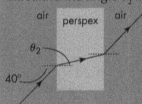

- First, use an appropriate equation: $n = \dfrac{\sin\theta_1}{\sin\theta_2}$, where '1' is the angle of incidence in the air.

$$1·51 = \left(\frac{\sin 40°}{\sin\theta_2}\right)$$

- Finally, solve for θ_2:

$$\theta_2 = \sin^{-1}\left(\frac{\sin 40°}{1·51}\right) = \underline{\underline{25°}}$$

4 Using the table of refractive indices find the angles labelled S, T, U, V, W and X.

Material	Refractive index (n)
Plate glass	1·52
Crown glass	1·54
Flint glass	1·60

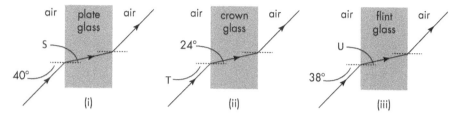

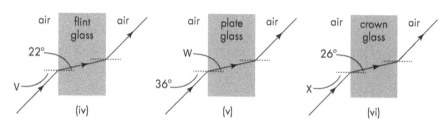

5 Find the missing angles labelled A, B, C, D, E, F, G and H in the diagrams below.

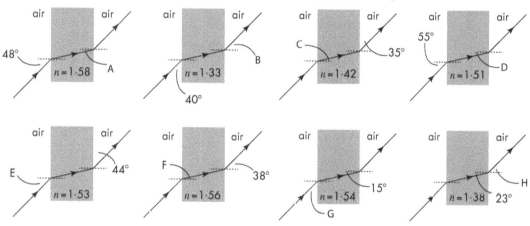

6 A student carries out some refraction experiments to identify the material that a beam of monochromatic light passes through. The angles of incidence and refraction are shown in the diagrams on page 92 and a table of refractive indices is shown below. Determine which diagram corresponds with which material.

Material	Refractive index (n)
Silica	1·46
Rocksalt	1·52
Pyrex	1·47
Acrylic glass	1·49
Cubic zirconia	2·17
Diamond	2·42
Amber	1·55
Water	1·33

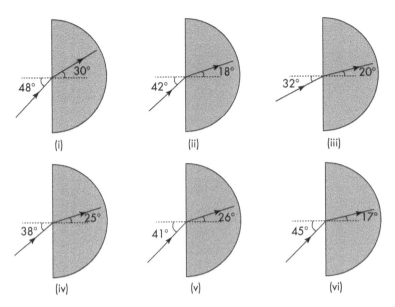

(i)

(ii)

(iii)

(iv)

(v)

(vi)

7 A ray of blue light ($\lambda = 500\,$nm) passes from air ($n_{blue} = 1.00$) into a glass block ($n_{blue} = 1.53$) as shown here.

Determine:

a the angle θ_2;

b the frequency of the light (i) in air and (ii) in the glass;

c the speed of the light in the glass;

d the wavelength of the light in the glass; and

e the region of the e-m spectrum where the wavelength calculated in part (d) above would lie.

For questions 8 → 22, **do not use** the refractive index tables above.

8 A ray of yellow monochromatic light with a wavelength of 590 nm in air enters into a block of Pyrex ($n = 1.47$). Calculate the wavelength and the colour of the light in the Pyrex.

9 A ray of monochromatic light enters into a block of water ice ($n = 1.31$) from the air. The wavelength of the light in the ice is 420 nm. Find the wavelength and colour of the light in the air. (You will need to use the data table to estimate the colour.)

10 The speed of a beam of monochromatic light in air is $3.00 \times 10^8\,$m s^{-1}. The beam enters into a block of cubic zirconia ($n = 2.17$). Find the speed of the light beam in the block.

11 A beam of monochromatic red light has a wavelength of 675 nm in air. It enters a block of glass, which causes its wavelength to decrease to 450 nm. Calculate its new speed in the block of glass.

12 A beam of monochromatic violet light has a wavelength of 420 nm inside a block of amber ($n = 1.55$). Find the wavelength of the light when the beam enters the air.

13 The wavelength of a beam of red light in air is 653 nm. When the beam of light enters into a beaker filled with kerosene its wavelength changes to 470 nm. Find the speed of the light beam in the kerosene.

14 The wavelength of a beam of monochromatic red light in air is 650 nm. When it enters water its speed reduces to $2.26 \times 10^8\,$m s^{-1}. Calculate its wavelength in the water.

15 A beam of monochromatic light enters a glass block from air at an angle of incidence equal to 40°. Calculate the speed of the beam of light in the glass block if the angle of refraction is 18·7°.

16 A beam of monochromatic light enters a slab of rocksalt from air at an incident angle of 33° to the normal. The beam has an angle of refraction equal to 21° inside the rocksalt. Calculate the speed of the beam in the rocksalt.

17 A beam of monochromatic light in air enters a beaker full of acetone with an angle of incidence equal to 23·5°. Its speed on entering the acetone is $2\cdot26 \times 10^8\,\text{m s}^{-1}$. Calculate its angle of refraction.

18 A beam of monochromatic light in air enters a beaker full of glycerol and has an angle of refraction equal to 26·5°. The speed of the light in the glycerol is $2\cdot04 \times 10^8\,\text{m s}^{-1}$. What is the angle of incidence at the air/glycerol boundary?

19 The wavelength of a beam of monochromatic red light is 620 nm in air. When it enters a glass of water its wavelength changes to 466 nm. If the angle of refraction is 18°, calculate its angle of incidence.

20 The wavelength of a monochromatic beam of red light in air is 640 nm. This beam is shone into a block of plate glass. The angle of incidence is 25° and the angle of refraction is 16°. Find the wavelength of the light in the glass block.

21 The wavelength of light inside a block of acrylic glass is 410 nm. The angle of incidence in air is 34° and the angle of refraction in the glass is 22°. Calculate the wavelength and colour of the light in the air.

22 A beam of red light that has a wavelength of 680 nm in air enters a block of silica with an angle of incidence equal to 36°. The wavelength of the light in the silica is 466 nm. Find the angle of refraction of the beam inside the silica block.

23 A technician carries out an experiment that involves measuring the angles of incidence in air and the corresponding angles of refraction of a beam of orange light in a beaker filled with a 20% glucose solution in water. Use the results to plot a graph of $\sin\theta_1$ against $\sin\theta_2$ and then use the graph to find the refractive index of the glucose solution.

Angle of incidence, θ_1 (°)	Angle of refraction, θ_2 (°)
20·0	14·6
30·0	21·6
40·0	28·2
50·0	34·3
60·0	39·6
70·0	43·7

24 Explain why white light splits up into the visible spectrum of colour when it is passed through a glass prism.

25 A beam of white light enters a triangular prism as shown here.

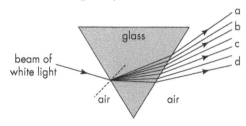

Identify the colours labelled a, b, c and d.

26 A glass block is used to pass monochromatic rays of red (a), green (b) and blue (c) light through the centre of the straight edge of the glass block shown. The refractive indices of the three colours in glass are shown in the table below.

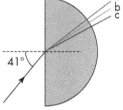

Colour	Refractive index (n)
Red	1·509
Green	1·515
Blue	1·517

a Why do the refracted coloured rays exit the semi-circular block without refracting?

b Calculate the angle of refraction in the glass block for each of the three colours used.

c Calculate the size of the angle between the red and the blue light.

27 A ray of white light passes through a triangular glass prism and splits into its component wavelengths. The paths taken by a yellow ray (a) and a violet ray (b) are shown.

The refractive index for yellow light in the prism is 1·511. The angle between the yellow ray and the violet ray when they first undergo refraction is 0·21°. Calculate the refractive index for the violet light in the glass prism.

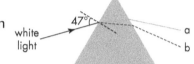

28 A ray of white light passes through a triangular glass prism and splits up as shown here.

The refractive index for green light in this glass is 1·515 and the refractive index for blue light is 1·517.

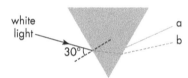

a Find the angle between the blue (a) and green (b) rays when they first refract on entering the prism.

b Using the refractive indices provided, determine which of the two colours is faster when travelling through the prism. Give an explanation for your answer.

29 A student refracts some green light through a semi-circular prism as shown.

a Calculate the refractive index for the green light in this glass.

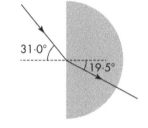

b Calculate the angle of refraction when the angle of incidence increases to 39°.

c Calculate the angle of incidence when the angle of refraction decreases to 14·5°.

30 A ray of light passes through four materials. The three materials present are water, glass and air. Given that $n_{glass} > n_{air} < n_{water}$, determine which material each of the rays $1 \rightarrow 4$ pass through.

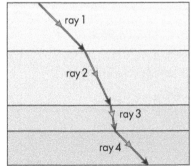

31 Explain how a rainbow is formed.

32 Describe two differences between a ray of white light passing through a double slit and a ray of white light passing through a glass prism.

33 Native Polynesians still fish by throwing a spear into the water. If the water is quite deep, where should the fisherman throw the spear, A, B or C? Explain.

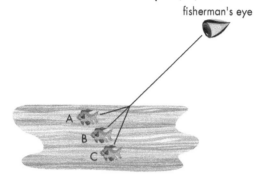

34 A ray of monochromatic red light passes from air into and out of an equilateral prism as shown. The refractive index of the glass used to make the prism is 1·51.

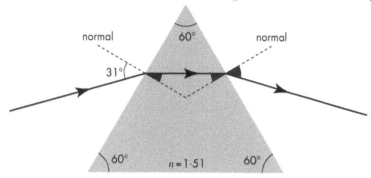

The angle of incidence is 31°. Find the shaded angles.

| Hint | If the triangle is equilateral, what angle will two normals make? |

Exercise 15B Critical angle and total internal reflection

1 What is meant by the 'critical angle' when applied to geometrical optics?

2 Define the term 'total internal reflection'.

3 Complete the following table:

Critical angle	Refractive index (n)
23°	
	1·55
41°	
	1·78
65°	
	2·50

4 The refractive index for crystal quartz is 1·549. A semi-circular crystal quartz block has two monochromatic beams of red light shone into it at 38° and 50° as shown.

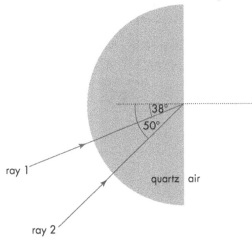

Draw the corresponding rays produced by each one when they strike the quartz–air boundary.

ray 1

quartz air

ray 2

5 A ray of monochromatic blue light passes from air into an amber block as shown.

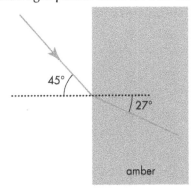

45°

27°

amber

What is the critical angle for the amber?

Example

A monochromatic beam of red light with a wavelength of 660 nm in air enters a glass of liquid with an angle of incidence equal to 41°. The wavelength of the light in the liquid is 428 nm.

a Calculate the critical angle of the liquid.

- Use an appropriate equation: $n = \dfrac{\lambda_1}{\lambda_2}$ where '1' is the wavelength in the air

$$= \frac{660 \times 10^{-9}}{428 \times 10^{-9}} = \underline{1·54}$$

b What is the angle of refraction in the liquid?

- Use an appropriate equation: $n = \dfrac{\sin\theta_1}{\sin\theta_2}$ where '1' is the angle of incidence in the air

$$1·54 = \frac{\sin 41°}{\sin\theta_2}$$

$$\theta_2 = \sin^{-1}\left(\frac{\sin 41°}{1·54}\right) = \underline{25°}$$

Hint For questions 6 → 9 refer to the table in question 6 in the previous exercise to help identify the material.

6 A monochromatic beam of red light with a wavelength of 630 nm in air enters a glass of liquid with an angle of incidence equal to 36°. The wavelength of the light in the liquid is 474 nm.

   a Calculate the critical angle of the liquid and also determine what liquid it is.

   b What is the angle of refraction in the liquid?

7 A monochromatic beam of green light travels from air into a material with an angle of incidence equal to 28°. The speed of the beam of light falls from $3.00 \times 10^8\,\mathrm{m\,s^{-1}}$ (in air) to $2.01 \times 10^8\,\mathrm{m\,s^{-1}}$ (in the material).

   a Calculate the critical angle of the material and also the material used.

   b What is the angle of refraction in the material?

8 A technician carries out an experiment where a monochromatic beam of red light passes from air into a slab of unknown material. The wavelength of the red light in air is 675 nm and the wavelength of the light in the material is 444 nm. The angle of refraction in the material is measured and found to be 28.2°.

   a Calculate the critical angle of the material and determine the material used.

   b Calculate the angle of incidence.

9 The speed of a monochromatic beam of red light inside a material is $1.24 \times 10^8\,\mathrm{m\,s^{-1}}$. The beam of light passes from air into the material and has an angle of refraction of 15.4°. Calculate the critical angle in the material and determine the material used.

10 One of the most appealing aspects of real diamonds is their ability to sparkle in natural daylight. Using your knowledge of physics, can you explain what it is about real diamonds that makes them sparkle so much more than 'fake' diamonds like cubic zirconia?

11 Rays of light emanate from the bottom of a 2.5 m deep pool of fresh water. When the rays reach the water–air boundary they form a circle of light on the surface with a radius 'r'. If the refractive index for water is 1.33 determine the radius of the circle of light on the surface.

Hint The rays from the lamp will refract along the water–air boundary when the angle of incidence equals the critical angle. Use the critical angle to determine 'r' with some simple trigonometry. Draw a sketch of the set-up.

Example

A ray of monochromatic light enters a triangular block as shown.

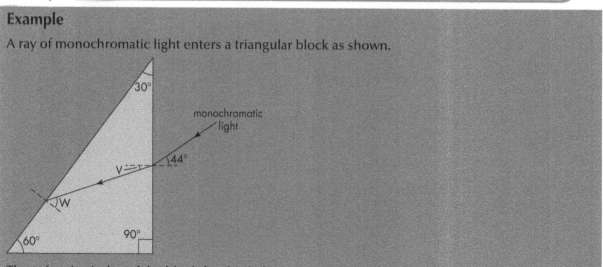

The refractive index of the block for this light is 1.39 and the angle of incidence is 44°.

a Calculate the size of angle V.

- First, use an appropriate equation: $n = \dfrac{\sin\theta_1}{\sin V}$ where '1' is the angle of incidence in the air

$$1{\cdot}39 = \frac{\sin 44°}{\sin V}$$

- Finally, solve for V: $\quad v = \sin^{-1}\dfrac{\sin 44°}{1{\cdot}39} = \underline{\underline{30°}}$

b Find the size of angle W.

- First, add together all the angles in the upper triangle formed to find the complimentary angle to W, i.e. $30° + 90° + 30° = 150°$
- Next, work out the complimentary angle: $180° - 150° = 30°$
- Finally, determine W: $90° - 30° = 60°$

c Complete the path of the ray of light until it emerges from the block. Include all important angles.

- First, calculate the critical angle by using the refractive index: $\quad \sin\theta_c = \dfrac{1}{n}$

$$\sin\theta_c = \frac{1}{1{\cdot}34}$$

$$\theta_c = \sin^{-1}\left(\frac{1}{1{\cdot}34}\right) = 48°$$

- Next, compare the angle of incidence to decide whether the ray undergoes refraction out of the block or undergoes total internal reflection. Since $W = 60°$ and is greater than the critical angle, the ray undergoes total internal reflection (see image).

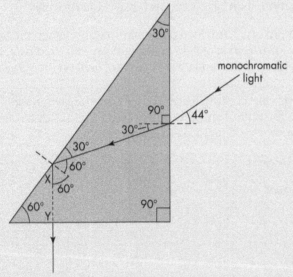

- Next, determine the values of X and Y: $\quad X = 90° - 60° = 30°$

$$Y = 180° - 60° - 30° = 90°$$

Since the angle of incidence is along the normal (i.e. 90°) the ray will pass straight out of the block.

12 A ray of monochromatic light enters a rectangular glass block as shown. The refractive index of the block is 1·52 and the angle of incidence is 52°.

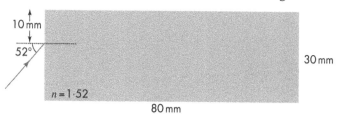

Copy the image exactly using the dimensions shown and complete the path of the ray of light including all angles until it leaves the block. Include the refracted angle when the ray leaves the block.

16 Spectra and model of the atom

Exercise 16A Spectra

Colour of light	Wavelength (nm)
Orange	590–640
Yellow	550–580
Green	490–530
Blue	460–480

1 Name five key features relating to the Bohr model of the atom.

2 What is a continuous spectrum and how is it formed?

3 What is a line emission spectrum and how is it formed?

4 What is a line absorption spectrum and how is it formed?

5 What kind of spectra would be observed by each of the following methods?

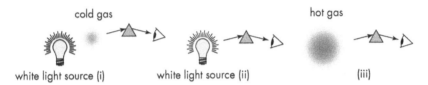

6 Different metal salts will produce colour when placed in the flame of a Bunsen burner. What causes the colours produced to be different?

7 What are Fraunhofer lines?

8 Identify the names of the various energy levels shown in the diagram.

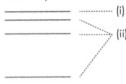

9 How many different photon energies are possible from the energy level diagram in question 8?

10 When line emission spectra are produced, do the electrons jump to higher or drop to lower energy levels?

11 When line absorption spectra are produced, do the electrons jump to higher or drop to lower energy levels?

Example

The energy level diagram shown here shows an electron dropping from E_4 to E_1.

- -8.71×10^{-20} J
- -1.36×10^{-19} J
- -2.42×10^{-19} J
- -5.42×10^{-19} J
- -21.8×10^{-19} J

Calculate the frequency of the photon emitted.

- First, calculate the energy difference between the two levels:

  $\Delta E = 5.42 \times 10^{-19} - 8.71 \times 10^{-20} = 4.55 \times 10^{-19}$ J

- Finally, use the appropriate equation to determine the frequency of the photon emitted:

  $\Delta E = hf$ where h = Planck's constant

  $4.55 \times 10^{-19} = 6.63 \times 10^{-34} \times f$

  $f = \underline{6.79 \times 10^{14} \text{ Hz}}$

12 In each of the energy level diagrams below calculate (a) the frequency of the photon emitted and (b) the wavelength of the photon emitted.

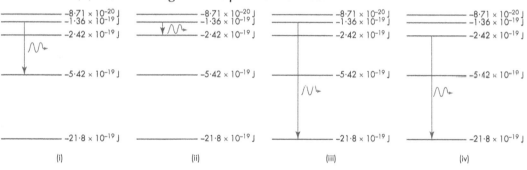

(i)   (ii)   (iii)   (iv)

13 Why do energy level diagrams show the energy state with a negative value?

14 The following apparatus is available to produce an absorption spectrum.

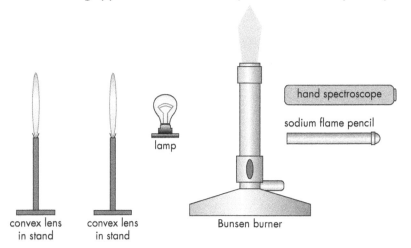

convex lens in stand   convex lens in stand   lamp   hand spectroscope   sodium flame pencil   Bunsen burner

Show the arrangement required to be able to produce the absorption spectrum of sodium.

15 During the production of an absorption spectrum, electrons gain energy that makes them transmit to higher energy states. These electrons however will quickly drop back down to lower energy states and re-emit the photons originally absorbed. Why are these photons not observed when the spectrum is analysed?

16 There are two very prominent lines in the line emission spectrum of sodium as shown.

700   600   500   400 nm

These are called the sodium-D lines. Why are these two lines more prominent than all the others in the line emission spectrum of sodium?

17 How does the line emission spectrum of elements allow them to be easily identified?

18 Electrons are accelerated in an x-ray tube by applying a p.d. of 90·0 kV. The electrons then collide with a heavy metal target causing electrons within the target atoms to excite into higher energy states. Excited electrons then quickly fall back down to lower energy states emitting x-ray photons in the process.

   a How much energy do the electrons absorb when they are accelerated within the x-ray tube?

   b What is the maximum energy available from the x-ray photons emitted?

   c What is the frequency of this photon?

   d What is the smallest wavelength photon that would be emitted by this x-ray tube?

19 One of the wavelengths of light emitted by a cadmium discharge tube has a wavelength of 643·8 nm. The time required to produce these photons by electron transitions is of the order of nanoseconds (ns). How many oscillations will occur during this transition?

> Hint   Consider the definition of frequency.

20 Shown here is a picture of part of the line emission spectrum for hydrogen together with its four lowest energy level states.

a Use the spectrum to help determine which transitions in the energy level diagram would give rise to the lines labelled (i) and (ii).

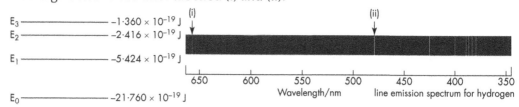

b Two transitions that take place are (i) $E_3 \to E_2$ and (ii) $E_1 \to E_0$.

For each one, determine whereabouts in the electromagnetic spectrum the photons emitted in each case would belong.

c Which transition gives rise to the photon with (i) the longest wavelength and (ii) the highest frequency?

d Which transitions will produce photons with wavelengths of (i) 102·8 nm and (ii) 97·5 nm?

e What is the minimum frequency of photon required to move an electron in the ground state to the ionisation level?

21 A photon is emitted from an infrared hydrogen laser when an electron makes the transition shown.

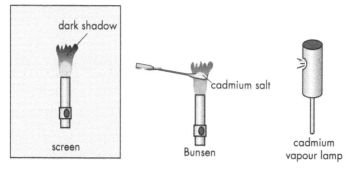

A pulse of light from this laser has an energy of 25 J. How many transitions are required to produce this pulse?

22 A cadmium vapour lamp is used to 'illuminate' a cadmium salt being vaporised in a Bunsen flame.

The image on the screen shows a dark shadow of the vaporised cadmium in the Bunsen flame.

a Explain why a dark shadow is observed on the screen.

b The cadmium vapour lamp is replaced by a sodium vapour lamp and the procedure repeated with the cadmium salt. Explain what happens to the dark shadow seen previously.

23 An electron with a kinetic energy of $1·52 \times 10^{-18}$ J collides with an electron in orbit around an atom, causing it to make a transition from an energy level with a value of $-9·93 \times 10^{-19}$ J to a higher energy level with a value of $-2·40 \times 10^{-19}$ J.

Calculate the final kinetic energy of the electron that makes the collision.

17 Measuring and monitoring alternating current

Exercise 17A Measuring and monitoring alternating current

1 What is the difference between 'peak' and 'r.m.s.' when referring to electrical signals?

2 Complete the table shown here:

r.m.s. voltage	Peak voltage
10·0 V	
	45·0 V
40 mV	
	7·0 mV
25 μV	
	100 μV

Example

An oscilloscope is used to monitor an a.c. signal produced by a signal generator.

a What is the peak voltage displayed?

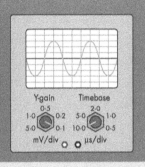

- First, measure the number of divisions vertically from the centre line to the peak (3 divisions here).
- Next, multiply this number of divisions vertically by the Y-gain reading on the CRO.

  $3 \times 0·1 \times 10^{-3} = \underline{0·3\,mV}$

b What is the frequency produced by the signal generator that provides this display?

- First, measure the number of divisions horizontally corresponding to 1λ (4 divisions here).
- Next, multiply this number of divisions horizontally by the timebase to find the period, T.

  $T = 4 \times 10·0 \times 10^{-6} = 40 \times 10^{-6}\,s$

- Finally, use an appropriate equation to find the frequency using the period T.

  $f = \dfrac{1}{T} = \dfrac{1}{40 \times 10^{-6}} = \underline{25\,kHz}$

For questions 3 and 4 below, the following key applies: 1 division [ ] 1 division

3 For each of the oscilloscope images below find the corresponding:

i peak voltage;

ii r.m.s. voltage; and

iii frequency of the signal.

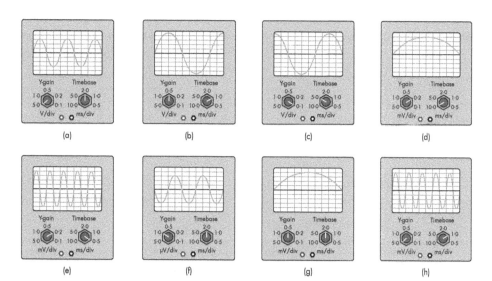

(a)   (b)   (c)   (d)

(e)   (f)   (g)   (h)

4 Use the oscilloscope trace and the corresponding output voltages and frequencies to find the Y-gain and timebase settings on each oscilloscope.

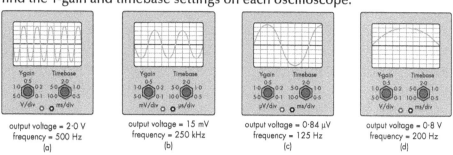

output voltage = 2·0 V
frequency = 500 Hz
(a)

output voltage = 15 mV
frequency = 250 kHz
(b)

output voltage = 0·84 µV
frequency = 125 Hz
(c)

output voltage = 0·8 V
frequency = 200 Hz
(d)

5 An a.c. circuit is set up and an oscilloscope monitors the output across a 560 Ω resistor as shown.

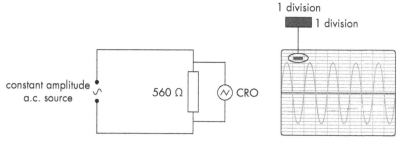

The CRO settings are as follows:

- Y-gain = 2 V/div

- Timebase setting = 5 m s/div

The CRO output is shown next to the circuit above.

a Calculate the frequency of the output from the a.c. supply.

b Calculate the r.m.s. voltage of the a.c. supply.

c Calculate the r.m.s. current in the 560 Ω resistor.

6 An oscilloscope is used to observe an output signal from a signal generator. Signal 1 is obtained when the oscilloscope settings are Y-gain = 5 mV/div and a timebase = 5 ms/div. The Y-gain and timebase are adjusted and signal 2 is shown on the oscilloscope.

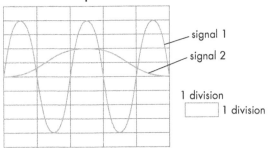

signal 1
signal 2

1 division
1 division

Calculate the new Y-gain and timebase settings of the oscilloscope required to produce signal 2 if the output from the signal generator remains unchanged.

Hint Signal 1 is $2\frac{1}{2}\lambda$ Signal 2 is 1λ.

7 The oscilloscope control settings are shown on the oscilloscope opposite.

a For this signal find (i) the peak voltage and (ii) the frequency.

b The timebase is now changed to 1·0 μs/div. Describe the signal now displayed by the oscilloscope.

c The timebase is returned to 2·0 μs/div and the Y-gain is changed to 0·1 mV/div. Describe the signal now displayed by the oscilloscope.

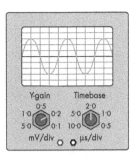

8 The output from an electronic circuit produces a display on an oscilloscope as shown.

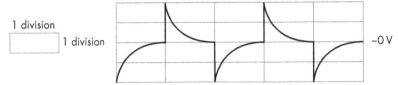

1 division
1 division
–0 V

The timebase setting of the oscilloscope is 0·5 ms/div and the Y-gain setting is 10 μV/div.

a Calculate the peak voltage output from the electronic circuit.

b Calculate the frequency of the electronic circuit.

9 A 12 V, 0·30 A lamp is connected to a 12 V d.c. battery. The ammeter reading is 0·30 A.

a Calculate the resistance of the lamp when it is operating normally.

b Calculate the power dissipated by the lamp when it is operating normally.

c The 12 V d.c. battery is now replaced by a 12 V a.c. supply.

   i Calculate the peak output voltage of the a.c. supply.

   ii Calculate the peak current in the filament of the lamp when it is operating normally.

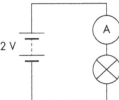

12 V

10 A student sets up a CRO together with a sound transmitter/receiver to measure the speed of an ultrasound pulse travelling through a 0·92 m copper bar. The pulse travels the length of the copper bar and then reflects back from the end of the bar towards the receiver.

The trace obtained by the CRO when the timebase control is set at 0·2 ms cm⁻¹ is shown below.

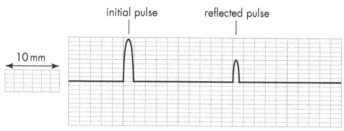

a Use the trace and the timebase setting to calculate the speed of sound in copper in m s⁻¹.

b Explain why the reflected pulse is smaller than the transmitted pulse.

18 Circuits: current, voltage, power and resistance

Exercise 18A Current, voltage, power and resistance

1 In each of the following circuits find the reading on the ammeters and/or voltmeters.

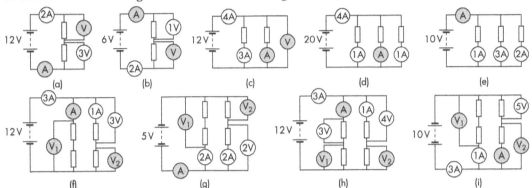

2 For each of the following circuits find (i) the total resistance and (ii) the total current.

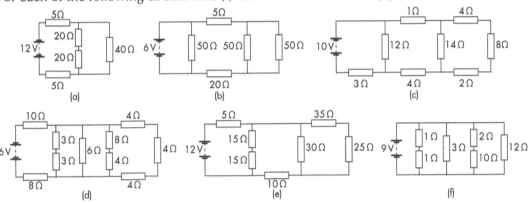

3 A technician uses three 3 kΩ resistors to construct a circuit which has a total resistance of 2 kΩ. The power supply used is 10 V.

   a Draw a circuit similar to that constructed by the technician.

   b Add another resistor to the circuit in order that its total resistance is now 1 kΩ.

   c Calculate the total current in the circuit.

   d Calculate the current through the resistor you added in part (b).

   e Calculate the current in each of the three 3 kΩ resistors.

4 Complete the following table:

Voltage (V)	Current (mA)	Resistance (Ω)	Power (mW)
12·0	10·0		
10·0			350
		250	10
25·0		125	
	2·50		6250

5 For each of the following circuits find the reading on the ammeters and voltmeters.

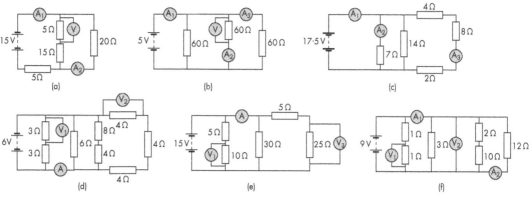

(a)　(b)　(c)

(d)　(e)　(f)

Example

A potential divider is set up as shown here.
What would be the reading on the voltmeter?

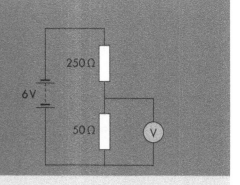

Use an appropriate formula: $V_2 = \left(\dfrac{R_2}{R_1 + R_2}\right)V_s$

$$V_2 = \left(\dfrac{50}{250 + 50}\right) \times 6 = \underline{1V}$$

6 Find the reading on the voltmeter in each of the following circuits:

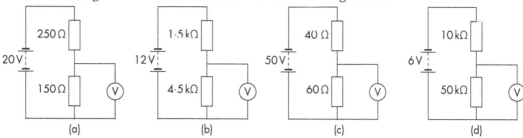

(a)　(b)　(c)　(d)

7 Find the value of R_2 in each of the circuits below:

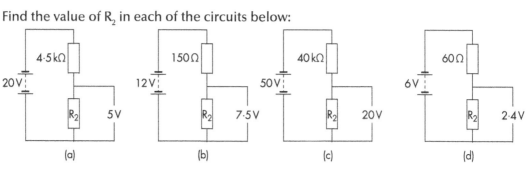

(a)　(b)　(c)　(d)

8 Find the value of R_1 in each of the circuits below:

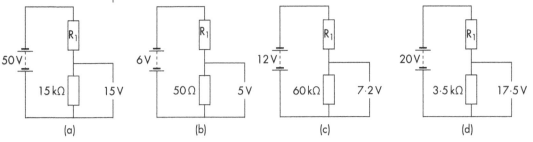

(a)   (b)   (c)   (d)

9 Find the value of V_s in each of the following circuits:

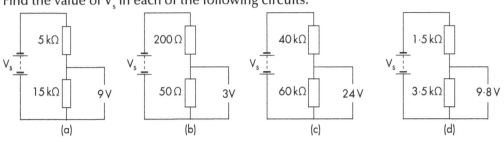

(a)   (b)   (c)   (d)

Example

Two potential dividers are set up as shown here.

What is the reading on the voltmeter?

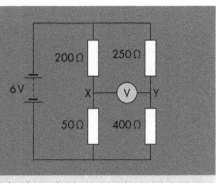

- First, ignore the two resistors on the right and concentrate on finding the potential at point X using the potential divider formula:

$$V_X = \left(\frac{R_2}{R_1 + R_2} \right) V_s$$

$$V_X = \left(\frac{50}{200 + 50} \right) \times 6 = 1 \cdot 2 \, V$$

- Next, ignore the two resistors on the left and concentrate on finding the potential at point Y using the potential divider formula:

$$V_Y = \left(\frac{R_2}{R_1 + R_2} \right) V_s$$

$$V_Y = \left(\frac{400}{250 + 400} \right) \times 6 = 3 \cdot 7 \, V$$

- Finally, the reading on the voltmeter will be the difference in these two voltmeter readings.

$$V = 3 \cdot 7 - 1 \cdot 2 = \underline{2 \cdot 5 \, V}$$

10 Each of the circuits below consists of two potential dividers connected in parallel.

a Find the reading on the voltmeter in each case:

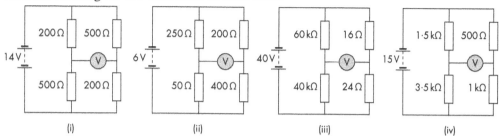

(i)   (ii)   (iii)   (iv)

b Does changing the supply voltage change the readings on the voltmeters? Explain.

11 Find the voltmeter reading for each of the following circuits:

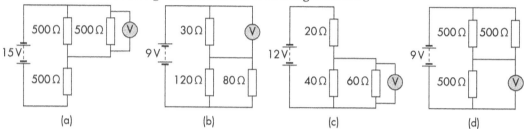

(a)   (b)   (c)   (d)

12 Calculate the reading on the voltmeter in the circuits below:

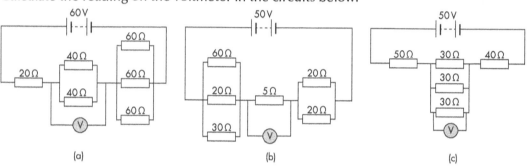

(a)   (b)   (c)

Example

A potential divider is set up as shown here.

a What is the reading on the voltmeter with
(a) switch 'S' open
and
(b) switch 'S' closed?

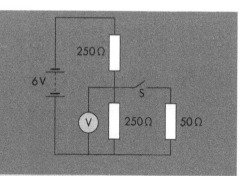

- First, use an appropriate formula: (a) with 'S' open

$$V = \left(\frac{R_2}{R_1 + R_2} \right) V_s$$

$$V = \left(\frac{250}{250 + 250} \right) \times 6 = \underline{3 \cdot 0\text{V}}$$

- Next,
  (b) with 'S' closed, calculate the total resistance of the parallel combination formed.

  $$\frac{1}{R_T} = \frac{1}{250} + \frac{1}{50} = 0{\cdot}024$$

  $$R_T = 42\,\Omega$$

- Next, use this value as R_2 in the potential divider formula.

  $$V = \left(\frac{R_2}{R_1 + R_2}\right)V_s$$

  $$V = \left(\frac{42}{250 + 42}\right) \times 6 = \underline{0{\cdot}86\,V}$$

13 For each of the circuits below, calculate the reading on the voltmeter when (i) the switch is open and (ii) the switch is closed.

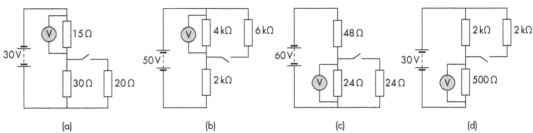

(a)   (b)   (c)   (d)

14 For each of the circuits below, calculate the reading on the voltmeter when (i) the switch is open and (ii) the switch is closed.

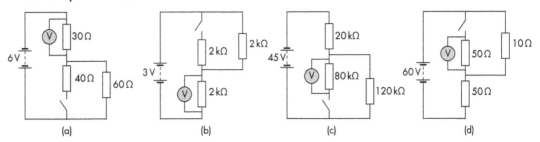

(a)   (b)   (c)   (d)

15 In the circuit shown here, the p.d. across the $60\,\Omega$ resistor is $40\,V$ when switch 'S' is open.

  a Find the supply voltage.

  b Calculate the p.d. across the $30\,\Omega$ resistor when switch 'S' is closed.

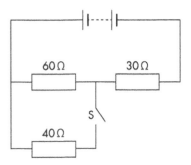

16 A student has six resistors to choose from – $2\,\Omega$, $5\,\Omega$, $10\,\Omega$, $30\,\Omega$, $50\,\Omega$, $100\,\Omega$ – together with a $12\,V$ power supply. Draw the potential divider arrangement that will provide an output of just $2\,V$.

17 A potential divider circuit is set up as shown in the circuit below. The lamp's normal operating parameters are 3 V, 0·03 A.

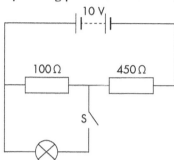

a What is the p.d. across the 450 Ω resistor when the switch 'S' is open?

b The switch 'S' is now closed. Does the lamp operate at normal full brightness? Explain.

c What size of resistor should replace the 450 Ω resistor so that the lamp operates at full brightness?

18 A light dependent resistor (LDR) makes up part of a potential divider circuit as shown here.

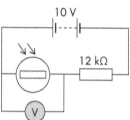

Condition	Resistance (Ω)
Daylight	400
Darkness	10×10^6

The resistance of the LDR in daylight and darkness is shown in the table. What will be the reading on the voltmeter in (a) daylight and (b) darkness?

19 Electrical sources and internal resistance

Exercise 19A Internal resistance

1. In the equation $E = V + Ir$ what do each of the symbols represent?

2. What is meant by the term 'e.m.f.' when applied to the equation above?

3. What is meant by the term 'lost volts' when applied to the equation above?

4. What is meant by the term 'terminal potential difference'?

5. What is a short-circuit current?

6. Why is a 12V car battery more dangerous when short-circuited than a 12V radio battery?

7. How should you measure the e.m.f. of a battery?

Example

A circuit is set up as shown here.

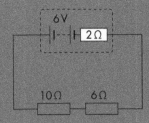

Calculate the circuit current, the t.p.d. and the lost volts.

- First, calculate the total external resistance, $R_T = 10 + 6 = 16\,\Omega$
- Next, use an appropriate formula: $E = IR + Ir$

  $E = I(R + r)$

  $6 = I(16 + 2)$

  $\underline{I = 0.33\,A}$

- Next, t.p.d., $V = IR = 0.33 \times 16 = \underline{5.3\,V}$
- Finally, lost volts $= Ir$

  or

  lost volts $=$ e.m.f. $-$ t.p.d. $= 6 - 5.3 = \underline{0.7\,V}$

  $0.33 \times 2 = \underline{0.7\,V}$

8 In each of the circuits below calculate (i) the circuit current; (ii) the t.p.d.; and (iii) the lost volts.

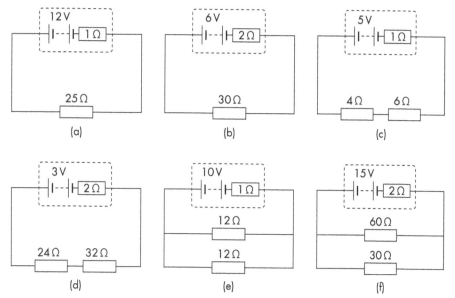

(a) (b) (c)

(d) (e) (f)

9 The circuit shown here shows a battery connected to an external fixed resistor, ammeter, voltmeter and switch.

The internal resistance 'r' is $2.50\,\Omega$. When the switch 'S' is open the voltmeter reads $12.0\,V$.

a State the e.m.f. of the battery.

b Calculate the reading on the ammeter when the switch is closed.

c Calculate the lost volts across the internal resistance when the switch is closed.

d Find the reading on the voltmeter when the switch is closed.

e What name is given to the voltage measured by the voltmeter when the switch is closed?

10 The circuit opposite is put together by a technician preparing apparatus for a lesson.

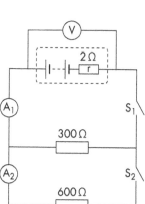

When switch 'S_1' is closed the voltmeter reading is $9.93\,V$ and the ammeter reading A_1 is $33\,mA$.

a Find the e.m.f. of the battery.

Switch 'S_2' is now also closed.

b Find the new reading on the ammeter A_1.

c Find the new reading on the voltmeter.

d Find the reading on ammeter A_2.

11 A battery has an e.m.f. of $15\,V$ and internal resistance of $1.5\,\Omega$. A current of $250\,mA$ is drawn from the battery by an external circuit.

a What is the resistance of the external circuit?

b What is the t.p.d.?

c How big is the lost volts?

12 The e.m.f. of a battery is 6 V. A current of 200 mA is drawn from the battery by an external circuit that has a total resistance of 28 Ω.

  a Find the internal resistance of the battery.

  b Find the lost volts.

  c Calculate the t.p.d.

13 The circuit opposite is set up using a 25 V power supply that has a 0·5 Ω internal resistance.

The initial external resistance is 12 Ω. The switch 'S' is now closed.

  a Find the current drawn from the supply by the external resistance.

  b Find the lost volts.

  c Calculate the t.p.d.

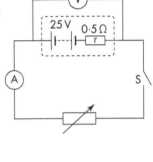

The external resistance is now increased and the current decreases to 1·2 A.

  d Find the value of this new external resistance.

  e What is the new lost volts?

  f Calculate the new t.p.d.

14 The circuit shown here is used to operate a motor.

The motor has a resistance of 50·0 Ω and the external variable resistor is set initially at 25·0 Ω.

  a What is the reading on the ammeter when switch 'S' is closed and the motor is operating?

  b Calculate the power dissipated in the motor.

  c What is the reading on the voltmeter when switch 'S' is closed?

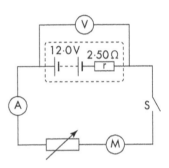

The variable resistance is now increased whilst the switch is closed.

  d Describe what happens to the ammeter reading.

  e What happens to the voltmeter reading? Justify your answer.

15 A circuit is set up with two resistors as shown.

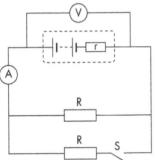

When the switch 'S' is closed what happens to the ammeter and voltmeter readings?

Justify your answers.

16 A 12 V car battery has an internal resistance of $1·0 \times 10^{-3}$ Ω. An auto-electrician accidentally shorts out the battery with a spanner whilst trying to remove it from the engine compartment.

How large is the short-circuit current?

19 Electrical sources and internal resistance

17 A student sets up a circuit to find the working parameters of a new battery. She draws the graph shown.

> **Hint** With regard to e.m.f. and internal resistance, look at the unit on the *y*-axis to see what the *y*-intercept is measured in. Once that's established, the gradient must be the 'other'.

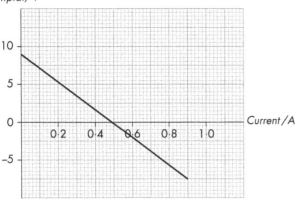

a Find the e.m.f. of the battery.

b Find the internal resistance of the battery.

c **Use the graph** to find the short-circuit current.

18 A teacher sets up a circuit and measures the t.p.d. of a battery as the current drawn from it is changed. The graph below is then plotted using the results.

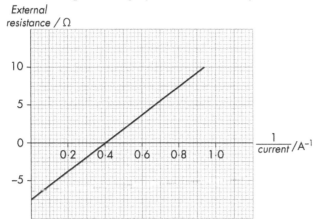

Use the graph to:

a find the e.m.f. of the battery; and

b find the internal resistance of the battery.

19 Three cells each with an e.m.f. of 1·5 V and internal resistance of 1·0 Ω are connected in series across an external resistance of 10·0 Ω.

a Calculate the t.p.d. when switch 'S' is closed.

b Calculate the current delivered by the three cells.

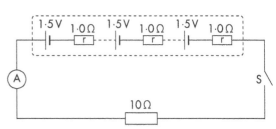

20 A battery has a high resistance voltmeter placed across its terminals. The voltage measured is 6·0 V with the switch 'S' open. The switch is now closed, connecting the battery to an external circuit that has a resistance of 24·0 Ω and causing the voltmeter reading to drop to 5·8 V.

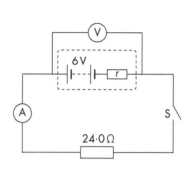

The ammeter reading when the switch closes is 230 mA.

  a What is the lost volts when current is drawn from the battery?

  b What is the internal resistance of the cell?

21 A circuit is set up by a school technician as shown.

When the switch 'S' is open the voltmeter reading is 4·5 V. When the switch is closed the voltmeter reading drops to 4·0 V and the ammeter reading is 600 mA.

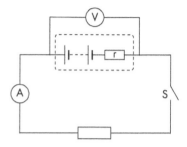

  a What is the internal resistance of the battery?

  b What is the external resistance of the circuit?

22 A car stalls at night at a set of traffic lights. The driver re-starts the car without switching off the headlights and notices that the car headlights dim as the car engine is re-starting. Explain this observation.

20 Capacitance

Exercise 20A Capacitance

1 Define the term 'capacitance'.

2 A student carries out an experiment using a capacitor that has a capacitance of $10\,\mu F$. What does 'a capacitance of $10\,\mu F$' actually mean?

3 Convert:

a $10\,pF$ to Farads (use scientific notation)

b $250\,nF$ to Farads (use scientific notation)

c $750\,\mu F$ to Farads (use scientific notation)

> **Hint** Learn how to use the (shift) ENG button on your calculator.

4 Convert:

a $0.0004\,F$ to mF (use scientific notation)

b $1.2 \times 10^{-5}\,F$ to μF

c $0.025\,F$ to mF (use scientific notation)

d $4.75\,F \times 10^{-4}$ to μF

e $0.0000025\,F$ to pF (use scientific notation)

f $0.125\,F$ to μF

g $3.3 \times 10^{-7}\,F$ to nF (use scientific notation)

h $1.2 \times 10^{-8}\,F$ to pF

5 A technician sets up a capacitor circuit using a power supply, a capacitor, a variable resistor, two voltmeters, an ammeter and two switches as shown.

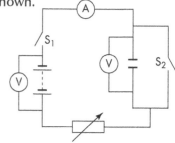

a How will the technician charge the capacitor?

b How will the technician know when the capacitor has been fully charged?

c How will the technician discharge the capacitor?

d How will the technician know when the capacitor has been fully discharged?

e If the technician wanted to charge the capacitor more quickly, using the same apparatus, what should she do?

Example

A circuit is set up to investigate the charging and discharging of a capacitor.

a Calculate the charge stored by the capacitor when switch S_1 is closed and the capacitor is fully charged.

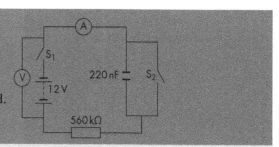

- Use an appropriate equation: $Q = VC$

$$= 12 \times 220 \times 10^{-9}$$

$$= \underline{2.6\,\mu C}$$

6 A student sets up a capacitor circuit using a 12·0 V power supply, a 220 μF capacitor, a variable resistor, an ammeter, voltmeter and two switches as shown.

Initially the variable resistor is set to 560 Ω. The student closes switch S_1 and allows the capacitor to charge.

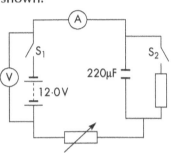

a Calculate the charge stored by the capacitor when it is fully charged.

b The variable resistor is now increased to 680 Ω. Calculate the charge stored by the capacitor now when it is fully charged.

Switch S_1 is now opened and switch S_2 is closed. The capacitor takes 12 ms to discharge.

c Calculate the average discharge current.

The 12 V power supply is now replaced by a 15 V power supply, S_2 is opened and S_1 is closed.

d Calculate the new charge stored by the capacitor.

7 A 560 μF capacitor is used to store 4·00 mC of charge. Calculate the power supply voltage used to charge the capacitor.

8 A capacitor is connected to a 10 V power supply and is fully charged. The charge stored is 500 nC. Calculate the size of capacitor used (in μF).

9 A 15 V power supply is used to charge a 750 nF capacitor. Calculate the charge stored by the capacitor in μC.

10 Complete the following table:

Charge	Average current	Time
25 mC		100 s
	50 mA	2 minutes
450 μC	9 μA	
	25 nA	5 minutes
0·02 C		20 s
75 μC	300 nA	

11 A student sets up the following capacitor circuit using a 9V power supply, a 50 µF capacitor and a 560 Ω fixed resistor.

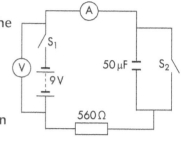

At a particular instant during charging the voltage across the capacitor is 4·0 V.

a Calculate the voltage across the resistor at that instant.

b What is the current when the voltage across the capacitor is 4·0V?

c How much energy (in µJ) is stored by the capacitor when the voltage across the capacitor is 4·0V?

12 A 250 µF capacitor is fully charged after 22 s using a 9V power supply.

a Draw a graph of charge stored by the capacitor against time. Include values on both axes.

b Draw a graph of voltage across the capacitor against time. Include values on both axes.

13 A 500 µF capacitor is fully charged by a 12V power supply. The capacitor is then fully discharged through a lamp in 6 s.

a Draw a graph of charge against time for the discharging capacitor. Include values on both axes.

b Draw a graph of voltage against time for the discharging capacitor. Include values on both axes.

14 A 560 pF capacitor is fully charged by a 9·0 V battery. Calculate the energy (in nJ) stored by the capacitor.

15 A 15V power supply is used to fully charge a capacitor. The charge stored is 400 mC.

a Calculate the energy stored by the capacitor.

The capacitor is now fully discharged through a lamp.

b What happens to the energy previously stored by the capacitor?

c Why does the capacitor not discharge immediately?

16 A 250 mF capacitor is fully charged and stores 500 µC of charge. Calculate the energy stored by the capacitor.

17 A 220 µF capacitor is charged by a 9V power supply by closing switch S_1.

The voltage across the resistor at one point during charging is 3·5 V.

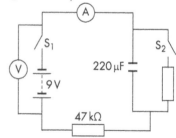

a Calculate the voltage across the capacitor when the voltage across the resistor is 3·5V.

A few seconds later the voltage across the capacitor is 7·5V.

b Find the current when the voltage across the capacitor is 7·5V.

Once the capacitor is fully charged switch S_1 is opened and switch S_2 is closed. At one instant during the discharge the voltage across the resistor is 5·0V.

c What is the voltage across the capacitor at the same instant?

18 A 120 µF capacitor is charged using a 12·0V power supply. At one instant during charging the voltage across the resistor, which is in series with the capacitor, is 4·0V.

Calculate the energy stored by the capacitor at that instant.

19 A capacitor is charged using a 15 V power supply. At one instant during charging the voltage across the resistor, which is in series with the capacitor, is 12 V and the energy stored by the capacitor is 7·5 mJ.

Calculate the charge stored by the capacitor at that instant.

20 A fully charged capacitor stores 100 mC of charge. When the capacitor is discharged through a resistor it releases 20 mJ of heat energy. What is the size of the capacitor used?

21 A capacitor is charged using a 6 V battery. The charging is stopped when the voltage across a resistor in series with the capacitor reaches 4 V. The capacitor is now discharged through a lamp releasing 3·96 μJ of energy. What is the size of the capacitor used?

22 A capacitor is being charged by a 15 V battery. The capacitor stores 400 mC of charge at an instant when 9 V is dropped across a resistor in series with the capacitor. Calculate the energy stored by the capacitor when fully charged.

23 A 120 mF capacitor is fully charged and stores 240 μC of charge. Calculate the energy stored by the capacitor.

24 A capacitor in series with a fixed value resistor is being charged by a 15 V battery. The capacitor stores 450 mC of charge and 1·125 J of energy at a particular instant during the charging.

Calculate the voltage across the resistor at that instant.

25 A 1000 nF capacitor in series with a fixed value resistor is being charged by a 9 V battery. At one instant during charging the energy stored by the capacitor is 4·5 μJ.

Find the voltage across the resistor at that instant.

26 A capacitor stores 400 mC of charge and 320 mJ of energy when fully charged.

Find the size of the capacitor in mF.

27 A 250 μF capacitor in series with a fixed value resistor is being charged by a 12 V battery. The capacitor stores 3·125 mJ of energy at a particular instant during the charging.

Calculate the voltage across the resistor at that instant.

28 A graph of the voltage dropped across a capacitor as it is charging is shown below.

a What is the voltage across the capacitor when it is fully charged?

b At what time is the voltage across the capacitor equal to 10 V?

c After exactly 4 s the capacitor has stored 2·2 mC of charge. What is the value of the capacitor used?

d How much energy is stored by the capacitor after exactly 4 s?

e After exactly 10 s approximately how much charge is stored by the capacitor?

f How much energy is stored by the capacitor now?

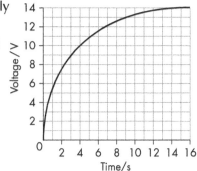

29 The circuit for a charging capacitor and its corresponding current–time graph is shown below.

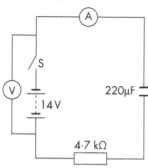 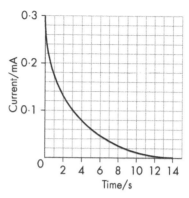

a How long does it take for the capacitor to charge when S_1 is closed?

b What is the maximum current when the capacitor starts to charge (do **not** use the graph)?

c What is the voltage across the resistor 4s into charging?

d What is the voltage across the capacitor 4s into charging?

30 A capacitor, resistor and ammeter are in series with a battery. A graph showing how the current in the ammeter varies with time is shown.

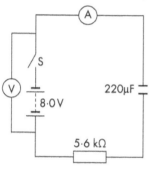 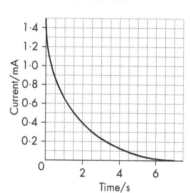

a 2s after switch S is closed, the voltage across the capacitor is 5·76 V. What is the size of the supply voltage used?

b How much energy is stored by the capacitor 2s after switch S is closed?

c How much charge is stored by the capacitor when the ammeter reads 1·07 mA?

d How much energy is stored by the capacitor when the ammeter reads 1·07 mA?

31 A voltage–time graph and a current–time graph for a small capacitor which is being charged are shown below.

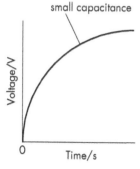 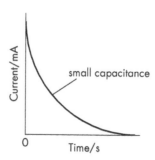

Copy the graphs and add the curves obtained when a capacitor of larger value is used.

32 A voltage–time graph and a current–time graph for a capacitor that is in series with a small resistor while it is being charged are shown below.

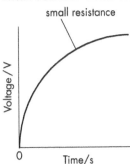

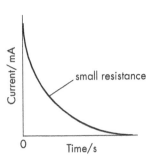

Copy the graphs and add the curves obtained when a resistor of larger value is used.

33 A capacitor connected to an a.c. supply and a fixed resistor is allowed to charge at different frequencies. The maximum current is measured at each frequency.

Sketch a current–frequency graph for the charging capacitor (no values are required).

34 A student carries out an experiment to charge a capacitor with different voltages and measure the corresponding charge acquired by the capacitor in each case using a coulombmeter. The results are shown in the table.

Voltage (V)	Charge (mC)
0·0	0
1·0	50
2·0	112
3·0	148
4·0	200
5·0	248
6·0	304

a Use these results to plot a charge–voltage graph.

b Use the graph to find the capacitance of the capacitor.

c Use the graph to find the total energy stored by the capacitor.

35 A 9V power supply is used to charge a capacitor using the circuit shown here.

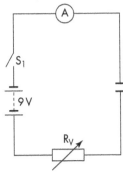

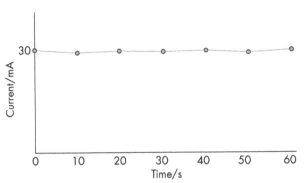

By using the variable resistor, the current is kept constant over a 60-second period. A graph of current against time is then plotted as shown.

a In what way is the resistance altered to produce the graph shown?

b Calculate the charge acquired by the capacitor over 60s.

c Calculate the size of the capacitor used.

36 The circuit shown here is used to charge a capacitor and measure the amount of charge stored using a coulombmeter.

The supply voltage used is a (9.0 ± 0.2) V d.c. battery. The switch is first thrown to position 1 to fully charge the capacitor and then to position 2 to discharge the capacitor through the coulombmeter. The charge/discharge procedure is carried out six times and the results are shown below.

2·01 mC   1·97 mC   2·02 mC
1·99 mC   1·98 mC   2·03 mC

a Calculate the mean charge measured by the coulombmeter.

b Calculate the approximate random uncertainty in the charge measured by the coulombmeter.

c Calculate the percentage uncertainty in both the battery voltage and the coulombmeter readings.

d Calculate the size of the capacitor used together with its **absolute** uncertainty. Leave your answer in the form: (capacitance ± absolute uncertainty).

two position switch

coulombmeter

9·0 V

47 kΩ

37 A defibrillator is being tested by a technician on a dummy which is manufactured from materials that have a similar resistance to human skin. The defibrillator has a 56 μF capacitor that is used to store charge before being administered to a person who is suffering from an irregular heartbeat using two paddles as shown.

The technician charges the capacitor using a 2·75 kV a.c. supply.

a Calculate the charge stored by the capacitor when it is fully charged.

b Calculate the maximum energy released when the capacitor discharges.

The material through which the current is released has a resistance of 90 Ω.

c Calculate the maximum discharge current that passes through the material between the paddles.

The material used is now changed to one with a smaller resistance of 75 Ω.

d How will this affect the time taken for the capacitor to discharge through the paddles?

e How will this affect the maximum discharge current?

f When used on a real patient the paddles usually have a lubricant gel applied prior to charging the capacitor. What purpose does this serve in administering the current to the patient?

38 A circuit used to charge and discharge a capacitor is shown below.

The two-position switch can be moved between S_1 and S_2. When in position S_1 the capacitor is charging and when in position S_2 the capacitor is discharging.

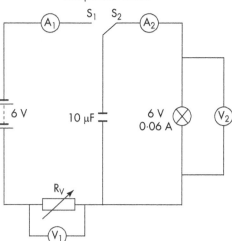

Initially, R_V is set to $200\,\Omega$ and S_1 is closed allowing the capacitor to charge fully.

a What is the maximum charging current measured by A_1?

b How much charge is transferred to the capacitor once fully charged?

c How much energy can the capacitor store?

After a few seconds ammeter A_1 reads $10\,\text{mA}$.

d What is the voltage across the resistor now?

e What is the voltage across the capacitor now?

f How much energy does the capacitor hold when A_1 reads $10\,\text{mA}$?

The switch is now moved to position S_2 and the capacitor starts to discharge through the lamp.

g What is the maximum discharging current measured by A_2?

h What is the maximum power dissipated by the lamp during discharge?

i What will happen to the lamp as the capacitor discharges?

j What will happen to the resistance of the lamp as the capacitor discharges? Explain.

k When the voltmeter registers $1.5\,\text{V}$ what is the voltage across the capacitor?

l How much energy is stored by the capacitor at this moment in time?

m What would R_V have to be set to in order to allow the time taken to charge to be the same as the time taken to discharge?

Once the capacitor is fully discharged R_V is now increased in value. The switch is moved back to position S_1 and the capacitor again allowed to charge.

n How will the increase in R_V affect the time taken to charge the capacitor?

o How will this increase in R_V affect the maximum current during charging?

p When the charge transferred reaches $20\,\mu\text{C}$ how much energy is stored by the capacitor?

The capacitor is again discharged and R_V is increased to $20\,\text{k}\Omega$. The capacitor is then charged and discharged over a $30\,\text{s}$ period as illustrated by the voltage–time graph below.

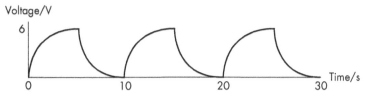

q Draw the corresponding current–time graph including values on both axes for the same period.

r What is the frequency of the switching used to produce the voltage–time graph shown above?

s Re-draw the above voltage–time graph if the frequency of switching is doubled.

21 Conductors, insulators and semiconductors / p–n junctions

Exercise 21A Conductors, insulators and semiconductors / p–n junctions

1 Complete the following statements using words from the item bank below.

Item bank:

full	small	not full	large

a In a semiconductor, the gap between the valence band and conduction band is _____

b In a conductor the valence band is _____

c In an insulator, the gap between the valence band and the conduction band is _____

d In an insulator the valence band is _____

2 Which of the following images represents a conductor, a semiconductor and an insulator?

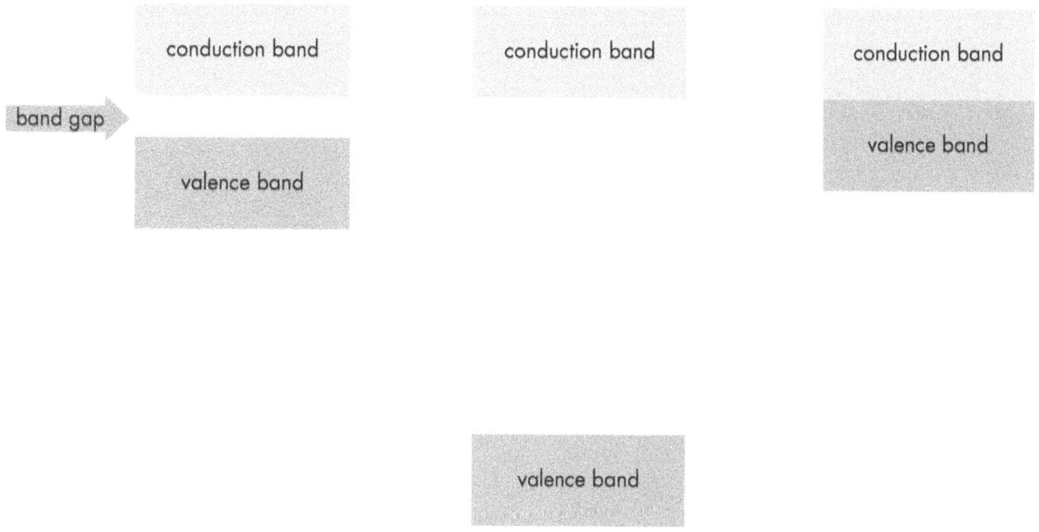

conduction band

band gap

valence band

conduction band

valence band

conduction band

valence band

(a)   (b)   (c)

3 Explain why pure silicon does not conduct.

4 Name two charge carriers used to describe semiconductor theory.

5 Describe a p-type semiconductor using the term 'doping'.

6 Describe an n-type semiconductor using the term 'doping'.

7 What happens to the overall charge of a semiconductor when doped? Explain.

8 Explain how a p–n junction diode can allow conduction to occur.

9 Why is a potential difference required to allow a p–n junction to conduct?

10 Describe the movement of the charge carriers when a large enough potential difference is applied across a p–n junction to allow the semiconductor to conduct.

11 Describe what is meant by a forward biased diode.

12 Draw a simple circuit showing a diode in forward bias mode.

13 Describe what is meant by a reverse biased diode.

14 Draw a simple circuit showing a diode in reverse bias mode.

15 Using band theory, explain how an LED operates and how coloured LEDs come about.

16 What is a solar cell and how does it work?

Leckie
the education publisher
for Scotland

Higher
PHYSICS

Mixed Exam
Question Practice

Paul Ferguson

MULTIPLE-CHOICE QUESTIONS

1 A student walks 6 km due east. The student then turns and walks 8 km due west. The total journey takes 3 hours.

Which row in the table shows the student's average speed and average velocity for the whole journey?

	Average speed	Average velocity
A	0·67 km h^{-1}	4·67 km h^{-1} due west (270)
B	0·67 km h^{-1}	0·67 km h^{-1} due east (090)
C	0·67 km h^{-1}	4·67 km h^{-1} due east (270)
D	4·67 km h^{-1}	0·67 km h^{-1} due west (270)
E	4·67 km h^{-1}	0·67 km h^{-1} due east (090)

2 A train of mass $7·0 \times 10^5$ kg is travelling at 55 m s^{-1} along a straight horizontal track. The brakes are applied and the train decelerates uniformly to rest in a time of 40 s.

The distance the train travels between the brakes being applied and the train coming to rest is

A 1100 m   **B** 2200 m   **C** 3300 m

D 3320 m   **E** 4400 m.

3 A block rests on a flat horizontal surface. A force of 50·0 N is now applied to the block and it moves horizontally to the left along the surface.

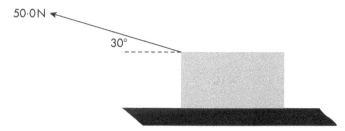

The mass of the block is 2·50 kg.

The acceleration of the block is 1·20 m s^{-2}.

The frictional force opposing the motion of the block is

A 22.0 N   **B** 28·0 N   **C** 40·3 N

D 43·3 N   **E** 46·3 N.

4 A vehicle of mass 1·0 kg moves from left to right along a horizontal frictionless air track. Another vehicle of mass 0·5 kg moves from right to left along the same track.

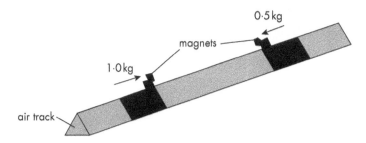

The vehicles collide and stick together using magnets.

Which of the following quantities is/are conserved in this collision?

1

i The kinetic energy

ii The total momentum

iii The total energy

A i only

B ii only

C iii only

D i and iii only

E ii and iii only

5 A planet of mass $8·52 \times 10^{26}$ kg orbits around a star of mass $4·02 \times 10^{30}$ kg. The star exerts a force of $1·03 \times 10^{29}$ N on the planet.

The mean distance of the star from the planet is

1

A $1·49 \times 10^{9}$ m

B $5·71 \times 10^{14}$ m

C $2·22 \times 10^{18}$ m

D $1·05 \times 10^{22}$ m

E $3·26 \times 10^{29}$ m.

6 An ambulance is travelling away from a stationary pedestrian at $22\,\text{m s}^{-1}$. The constant frequency emitted by the ambulance siren is $700\,\text{Hz}$. The speed of sound in air is $343\,\text{m s}^{-1}$.

The frequency of the ambulance siren heard by the pedestrian is

A   655 Hz

B   658 Hz

C   700 Hz

D   745 Hz

E   748 Hz.

1

7 A hadron called a lambda particle consists of three different quarks (uds). Information relating to the charge on each quark is shown below.

Quark	Charge
up (u)	$+\frac{2}{3}$
down (d)	$-\frac{1}{3}$
strange (s)	$-\frac{1}{3}$

Which row in the table below shows the type of hadron and the charge of the lambda particle?

	Type of hadron	Charge
A	meson	$+\frac{4}{3}$
B	meson	$-\frac{4}{3}$
C	meson	0
D	baryon	0
E	baryon	$+\frac{2}{3}$

1

8 Hubble's law allows physicists to estimate

A   the temperature of stellar objects

B   the universal constant of gravitation, G

C   the amount of dark matter in the universe

D   the mass of a galaxy using the orbital speed of stars within the galaxy

E   the age of the universe.

1

9 A student makes the following statements about the expanding universe.

i Hotter objects emit less radiation per unit surface area per unit time than cooler objects.

ii The peak wavelength of emitted radiation from a stellar object is shorter for hotter objects than for cooler objects.

iii Some of the evidence supporting the Big Bang and subsequent expansion of the universe comes from the abundance of the elements hydrogen and helium in the cosmos.

Which of these statements is/are correct?

A i only

B ii only

C i and ii only

D ii and iii only

E i, ii and iii

1

10 The period T of a planet around a star in a circular orbit is given by the relationship

$$T = \sqrt{\frac{4\pi^2 r^3}{Gm_s}}$$

where,

m_s is the mass of the star in kg

G is the universal constant of gravitation in $m^3\,kg^{-1}\,s^{-2}$

r is the radius of the orbit of the planet around the star in m.

A planet of mass $7 \cdot 22 \times 10^{25}$ kg is orbiting a star of mass $2 \cdot 63 \times 10^{30}$ kg.

The period of rotation of the planet around the star is $4 \cdot 73 \times 10^6$ s.

The radius of orbit of the planet around the star is

1

A $2 \cdot 76 \times 10^6$ m

B $1 \cdot 40 \times 10^9$ m

C $4 \cdot 63 \times 10^{10}$ m

D $9 \cdot 97 \times 10^{15}$ m

E $9 \cdot 94 \times 10^{31}$ m.

11 A particle with a charge of −6·0 mC is released from rest in an electric field between two parallel metal plates. The particle has a mass of $2·3 \times 10^{-11}$ kg.

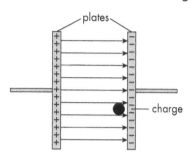

The potential difference between the plates is 1·5 kV.

The speed of the particle on reaching the positive plate is   1

A $2·0 \times 10^4\,\mathrm{m\,s^{-1}}$

B $6·3 \times 10^5\,\mathrm{m\,s^{-1}}$

C $8·9 \times 10^5\,\mathrm{m\,s^{-1}}$

D $2·8 \times 10^6\,\mathrm{m\,s^{-1}}$

E $7·8 \times 10^{11}\,\mathrm{m\,s^{-1}}$.

12 An electron enters a magnetic field as shown.

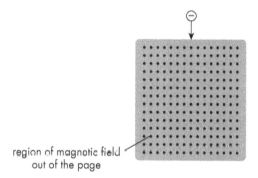

region of magnetic field
out of the page

On entering the magnetic field, the electron   1

A   deflects to the left

B   deflects to the right

C   deflects out of the page

D   deflects into the page

E   passes straight through undeflected

13 The star Sirius is 81.46×10^{12} km from Earth.

The star Betelgeuse is 6.08×10^{15} km from Earth.

How many orders of magnitude greater is the distance of the star Betelgeuse compared to Sirius?

  A 1 order

  B 2 orders

  C 3 orders

  D 4 orders

  E 5 orders

14 In a linear accelerator

  A the magnetic field is used to accelerate the charged particles.

  B the electric field is used to accelerate the charged particles.

  C the magnetic field accelerates the charged particles and the electric field is used to change the direction of the charged particles.

  D the electric field is used to change the direction of the charged particles and the magnetic field is used to change the direction of the charged particles.

  E there is no acceleration of the charged particles but the magnetic field changes the direction of the charged particles.

15 A student makes the following statements about the model of the atom.

  i   The atomic number is the number of neutrons in an atom.

  ii The mass number is equal to the number of protons and neutrons in an atom.

  iii An isotope has the same number of neutrons but a different number of protons.

Which of these statements is/are correct?

  A i only

  B ii only

  C iii only

  D i and iii only

  E ii and iii only

16 The following equation represents a fission reaction that occurs in a nuclear reactor.

$$^{X}_{92}U \ + \ ^{1}_{0}n \ \longrightarrow \ ^{144}_{Y}Ba \ + \ ^{89}_{36}Kr \ + \ 3^{1}_{0}n \ + \ energy$$

Which row in the table shows the values of X and Y?

	X	Y
A	233	53
B	234	56
C	234	53
D	235	56
E	235	53

17 A student makes the following statements about spectra.

i A line emission spectrum consists of black lines on a continuous coloured background.

ii An absorption spectrum can be used to determine the elements present in the outer atmosphere of a star.

iii A prism can produce a continuous spectrum by diffraction of white light.

Which of these statements is/are correct?

A i only

B ii only

C i and ii only

D ii and iii only

E i, ii and iii

18 A ray of red light with a frequency of $4 \cdot 60 \times 10^{14}$ Hz in air passes into a diamond.

Which row in the table gives the correct values for the speed, frequency and wavelength of this light in the diamond?

	Velocity (ms^{-1})	Frequency (Hz)	Wavelength (m)
A	$1 \cdot 24 \times 10^8$	$1 \cdot 92 \times 10^{14}$	$6 \cdot 46 \times 10^{-7}$
B	$1 \cdot 24 \times 10^8$	$4 \cdot 60 \times 10^{14}$	$2 \cdot 69 \times 10^{-7}$
C	$3 \cdot 00 \times 10^8$	$1 \cdot 92 \times 10^{14}$	$15 \cdot 6 \times 10^{-7}$
D	$3 \cdot 00 \times 10^8$	$1 \cdot 92 \times 10^{14}$	$2 \cdot 69 \times 10^{-7}$
E	$3 \cdot 00 \times 10^8$	$4 \cdot 60 \times 10^{14}$	$6 \cdot 46 \times 10^{-7}$

19 The diagram shows the path of a ray of green light passing from air into a transparent material.

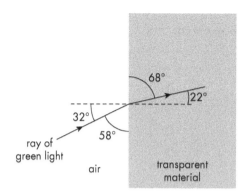

The critical angle for this light in the transparent material is

1

A   24°

B   26°

C   35°

D   45°

E   66°.

20 A student sets up the following experiment in a dark room to examine the inverse square law using a white point light source.

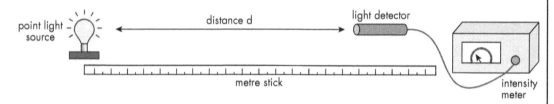

The initial irradiance of the light source at a distance of 0·80 m from the source is 0·064 Wm⁻².

The student now moves the light detector until the intensity meter reads 1·000 Wm⁻².

The distance from the light source to the detector is now

1

A   0·02 m

B   0·04 m

C   0·05 m

D   0·20 m

E   0·22 m.

21 A student sets up the circuit shown.

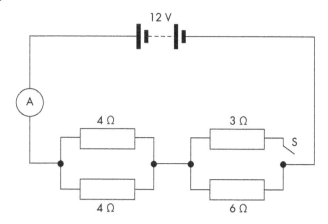

Which row in the table shows the ammeter reading when the switch S is open and when the switch S is closed?

1

	S open (A)	S closed (A)
A	0·7	0·3
B	0·9	0·7
C	1·2	2·9
D	1·5	3·0
E	6·0	3·0

22 A technician set up the following circuit for a classroom lesson.

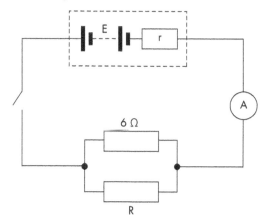

With the switch open, the emf E is 6·0 V.

The internal resistance r is 2Ω.

With the switch closed, the reading on the ammeter is 1·0 A.

The resistance of R is

1

A 2Ω **B** 3Ω **C** 6Ω

D 8Ω **E** 12Ω.

23 The following capacitor circuit is used to charge and discharge a capacitor.

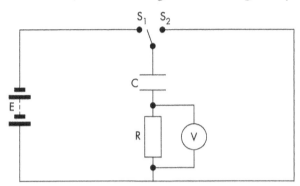

When the switch is placed in position 1, the capacitor charges through resistor R.

When the switch is placed in position 2, the capacitor discharges through resistor R.

Which of the following graphs shows how the voltage varies with time as the capacitor charges and discharges?

1

A

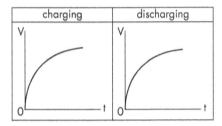

B

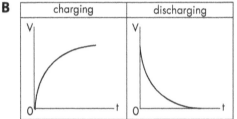

C

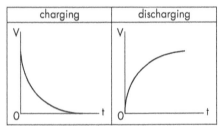

D

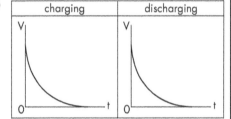

E

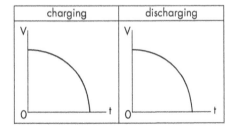

24 A student makes the following statements about semiconductors.

i Solar cells are p–n junctions designed so that a potential difference is produced when photons are lost.

ii LEDs are forward-biased p–n junction diodes that emit photons.

iii Doping a silicon semiconductor with specific impurities will increase the resistance of the semiconductor.

Which of these statements is/are correct?

A i only

B ii only

C iii only

D i and ii only

E ii and iii only

25 Data regarding the time taken for a capacitor to fully charge several times is shown below.

| 22·5 s | 22·0 s | 22·5 s | 21·5 s | 23·0 s | 22·5 s |

The approximate random uncertainty in the mean value of the time is

A 0·17 s

B 0·20 s

C 0·25 s

D 0·30 s

E 0·33 s.

EXTENDED RESPONSE QUESTIONS

1 A ball is fired vertically upwards with an initial velocity of 4·01 ms⁻¹ from a projectile launcher moving horizontally at 0·50 ms⁻¹ along a straight level bench.

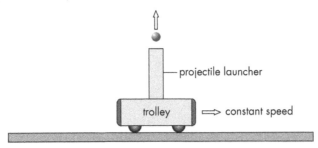

a Calculate the time taken for the ball to reach its highest height.

Space for working and answer

3

b Determine the magnitude of the velocity of the ball after 0·5 s.

Space for working and answer

5

c Determine the height of the ball after 0·5 s.

Space for working and answer

3

Total marks 11

2 A light canopy containing lamps is hung above a pool table as shown.

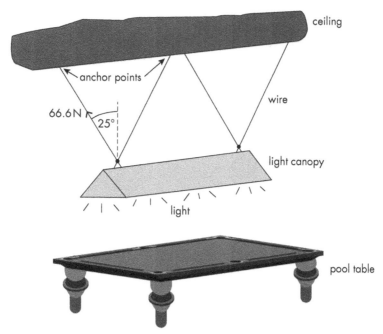

The canopy is connected to the ceiling by 4 wires of negligible mass, attached to the ceiling using anchor points. Each wire is 2 m long.

The tension in each wire is 66·6 N at an angle of 25° to the vertical.

a Calculate the vertical component of the force exerted by a single wire.

3

Space for working and answer

b Determine the weight of the canopy and the lamps.

2

Space for working and answer

c During refurbishment, the angle between the wires and the vertical is increased. The length of the wires does not change.

Explain what happens to the tension in each wire.

4

When the anchor points are separated, the height of the canopy above the pool table increases.

d Determine the change in the height of the canopy above the pool table when the angle increases from 25° to 30°.

3

Space for working and answer

Total marks 12

3 Sometimes, teachers use an analogy of a person walking up and down a ladder to describe the electron transitions between energy levels in an atom.

Use your knowledge of Physics to comment on this analogy.

3

4 The line absorption spectra of a particular element from two different galaxies is compared with a laboratory source of the same element on Earth.

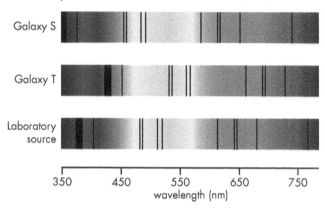

a Describe, in terms of electron energy level transitions, how the line absorption spectra are produced.

2

b Identify which galaxy is travelling towards Earth. Justify your answer.

3

c The wavelength of an absorption line from the laboratory source is measured to be 640 nm.

The same absorption line from Galaxy T is observed to have a wavelength of 690 nm.

Calculate the red shift in the light from Galaxy T.

3

Space for working and answer

d Calculate the speed of Galaxy T relative to the Earth.

3

e Calculate the approximate distance to Galaxy T.

3

Space for working and answer

Total marks | **14**

Extended response questions

5 The Standard Model is a model of fundamental particles and their interactions.

The following diagram provides some information about fundamental particles.

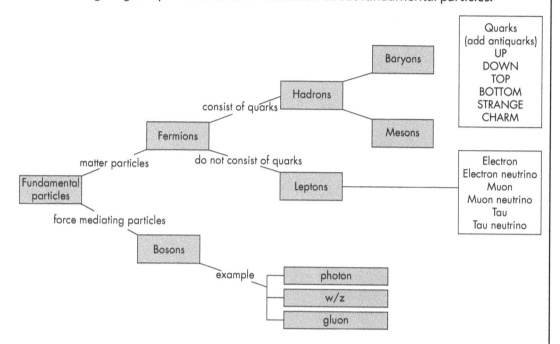

a Explain why hadrons are fermions but fermions are not necessarily hadrons.

1

b i State the number of quarks that make up a baryon.

1

ii State the number of quarks that make up a meson.

1

c What is the force associated with the w/z particle?

1

d A data table of the charge associated with quarks is shown below.

Quark	Symbol	Charge (e)
Up	u	$+\dfrac{2}{3}$
Down	d	$-\dfrac{1}{3}$
Top	t	$+\dfrac{2}{3}$
Bottom	b	$-\dfrac{1}{3}$
Charm	c	$+\dfrac{2}{3}$
Strange	s	$-\dfrac{1}{3}$

One particular particle is called a charmed sigma particle and is a baryon composed of only down and charm quarks. The charged sigma particle has an overall charge of zero. Identify the quarks that make up a charged sigma particle.

You must justify your answer.

Space for working and answer

2

Total marks **6**

6 A linear accelerator is used to accelerate protons to very high speeds.

Part of a linear accelerator is shown below.

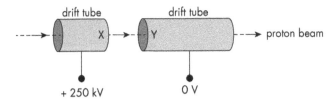

The proton beam passes through a series of drift tubes that increase in length.

A potential difference of 250 kV is maintained between each pair of drift tubes.

a What is meant by a *potential difference of 250 kV?*

1

b The protons accelerate in the gaps between each pair of drift tubes.

The speed of the protons as they leave the tube at point X is $2 \cdot 30 \times 10^6 \, ms^{-1}$.

Calculate the speed of the protons as they reach the next adjacent tube at point Y.

5

Space for working and answer

c Why does the length of the drift tubes increase when the gap between the tubes remains constant?

3

d As the protons enter a drift tube, the potential of all the tubes reverses. Why is this necessary?

1

e The gap between the drift tubes is doubled. Explain what happens to the speed of the protons at point Y.

2

Total marks 12

7 When explaining the Photoelectric Effect, the following equation is often used by teachers:

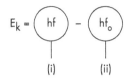

$$E_k = \underset{(i)}{\underbrace{\left(hf \right)}} - \underset{(ii)}{\underbrace{\left(hf_o \right)}}$$

Three parts of this equation represent three different terms used in explaining the Photoelectric Effect.

a State the meaning of the terms (i) and (ii).

i _____ 1

_____

ii _____ 1

_____

b An experiment involves shining different light sources onto the surface of a metal. The frequency of light ranges from 3.8×10^{14} Hz to 12.0×10^{14} Hz. The following graph is produced.

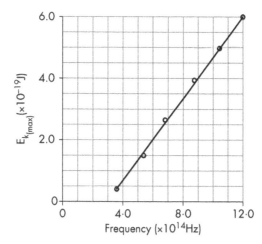

Frequency ($\times 10^{14}$ Hz)

The equation $E_k = hf - hf_0$ can be compared to the mathematical equation for a straight line graph, $y = mx + c$ as follows:

$$E_k = h \quad f \quad -hf_0$$
$$y = m \quad x \quad +c$$

i Use the graph to determine the threshold frequency of the metal used.

Space for working and answer

2

ii Use the graph to determine the value of Planck's constant.

Space for working and answer

3

c Using the value of Planck's constant determined above, calculate

   i the value of hf_0 **2**

     Space for working and answer

   ii the maximum kinetic energy of an escaping electron when the source frequency used is $3{\cdot}8 \times 10^{14}$ Hz.   **2**

     Space for working and answer

d Explain why a light source with a wavelength of $1{\cdot}5 \times 10^{-6}$ m was not used in the experiment.

  You must justify your answer.   **4**

Total marks **15**

8 Two experiments involving microwaves are carried out by a student.

The first experiment involves using a microwave transmitter to investigate interference.

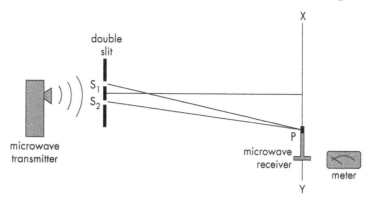

The microwaves pass through a double slit and are then detected using a detector that moves back and forth in a straight line between points X and Y. A meter that is connected to the detector indicates when a maximum or minimum is detected.

The path length from the slit S_1 to the detector at point P is 0·309 m.

The path length from the slit S_2 to the detector at point P is 0·302 m.

The wavelength of the microwaves used is 2·8 mm.

a Calculate the order number of the minimum detected at point P.

Space for working and answer

3

b What would happen to the position of point P along the line XY if the spacing between S_1 and S_2 is increased?

You must justify your answer.

4

The second experiment involves the microwaves passing through a narrow single slit into a wax prism that has a refractive index of 1·45.

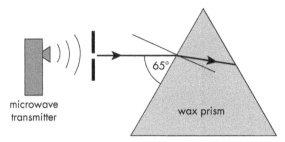

microwave
transmitter

65°

wax prism

The angle between the microwave beam and the face of the prism is 65°.

c   Calculate the angle of refraction of the microwave beam inside the wax prism.

Space for working and answer

3

d   Calculate the speed of the microwaves inside the wax prism.

Space for working and answer

3

e   What is the frequency of the microwaves inside the wax prism?

Space for working and answer

3

Total marks 16

9 A student sets up a circuit to investigate how the change in external resistance affects the current in the circuit.

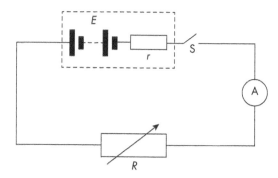

The following data is recorded:

External resistance R (Ω)	Current I (A)
1·00	2·00
2·00	1·50
3·00	1·20
4·00	1·00
5·00	0·86
6·00	0·75

a Calculate the value of $\frac{1}{I}$ for each external resistance, R.

External resistance R (Ω)	Current 1/I (A⁻¹)
1·00	
2·00	
3·00	
4·00	
5·00	
6·00	

1

b Using the graph paper provided at the end of this book, plot a graph of R against $\frac{1}{I}$ using axes laid out as follows:

3

The following relationship can be applied to the circuit:

$$E = I(R + r)$$

c Show that this relationship can be re-written as follows:

$$R = \frac{E}{I} - r$$

Space for working and answer

2

d Using the graph, determine

i the emf, E of the power supply.

Space for working and answer

2

ii the internal resistance, r of the power supply.

Space for working and answer

2

e Explain why the terminal potential difference, V increases as the external resistance increases.

4

Total marks 14

(10) The use of analogies from everyday life can help improve the understanding of physics concepts.

Water flowing in pipes is considered a useful analogy for electricity flowing in a circuit.

Use your knowledge of Physics to comment on this analogy.

3

11 A technician sets up a circuit as shown.

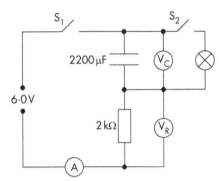

The capacitor is initially uncharged and switches S_1 and S_2 are both open.

Switch S_1 is now closed and the capacitor begins to charge.

a After a short period of time, the voltage reading on V_C is 1·8V.

Calculate the reading on the ammeter at this instant.

Space for working and answer

4

The capacitor is fully charged and switch S_1 is opened.

b Switch S_2 is now closed and the capacitor discharges through the lamp in 0·22 s.

Calculate the charge flowing in the lamp during the discharge of the capacitor.

Space for working and answer

3

c Calculate the current in the lamp during discharge.

Space for working and answer

d What is the power dissipated by the lamp?

Space for working and answer

Once the capacitor is fully discharged, switch S_2 is opened.

The 2 kΩ resistor is now replaced with a 10 kΩ resistor and again charged.

e What happens to:

i the maximum energy stored by the capacitor?

You must justify your answer.

ii   the time taken to fully charge the capacitor?

You must justify your answer.

Total marks 18

12   Data regarding some moons orbiting around a planet is shown below:

Period ($\times 10^3$ s)	Mean Orbital Radius ($\times 10^6$ m)
81·2	185
118	237
163	294
175	309
236	377

The relationship between the mean orbital radius and the orbital period of a moon about its planet is given by

$$R^3 = \frac{GM}{4\pi^2} T^2$$

where, R is the mean orbital radius in metres of the moon around its planet;

M is the mass of the planet in kilograms; and

T is the orbital period of the moon in seconds around its planet.

a   Calculate the value of T^2 and R^3 for each moon.

T^2 (s^2)	R^3 (m^3)

5

b Using the graph paper provided at the end of this book, draw a graph of R^3 against T^2.

3

c Find the gradient of your graph.

2

Space for working and answer

d Use your gradient to find an approximate value for the mass of the planet, M.

3

Space for working and answer

e The planet in part (d) has a mean orbital radius of 6.24×10^8 m and a mass of 1.14×10^{23} kg.

Calculate the period of rotation of this planet around its sun.

3

Space for working and answer

Total marks 16

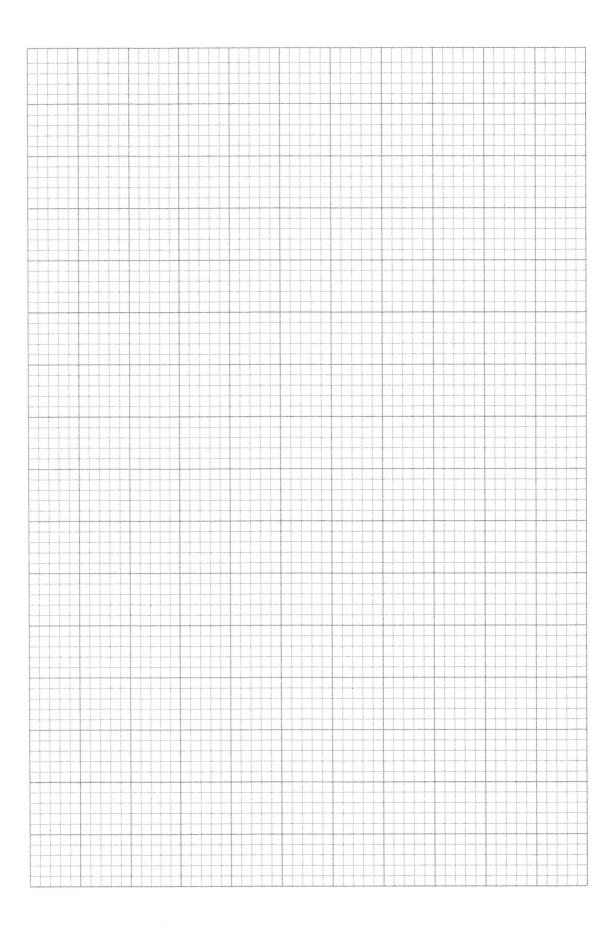

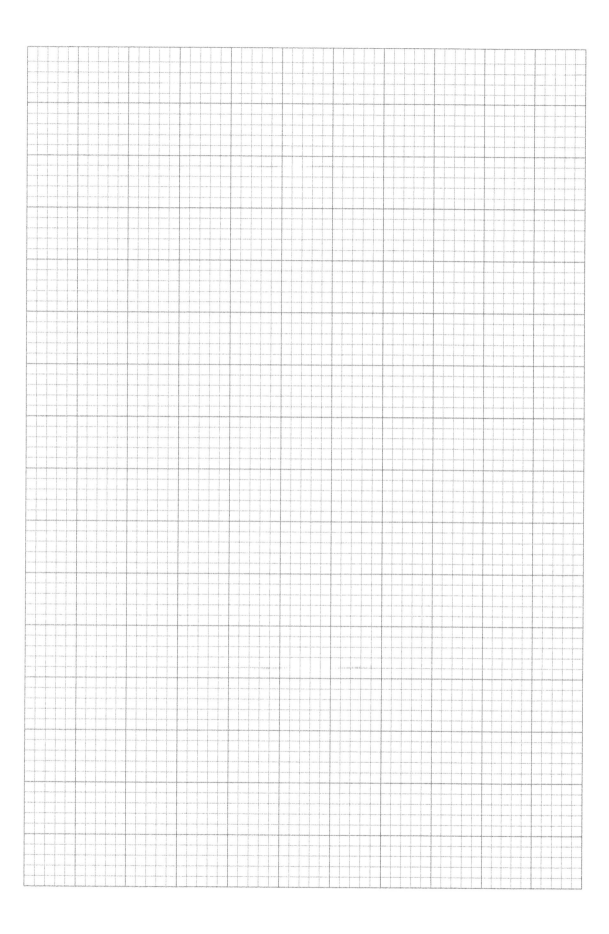

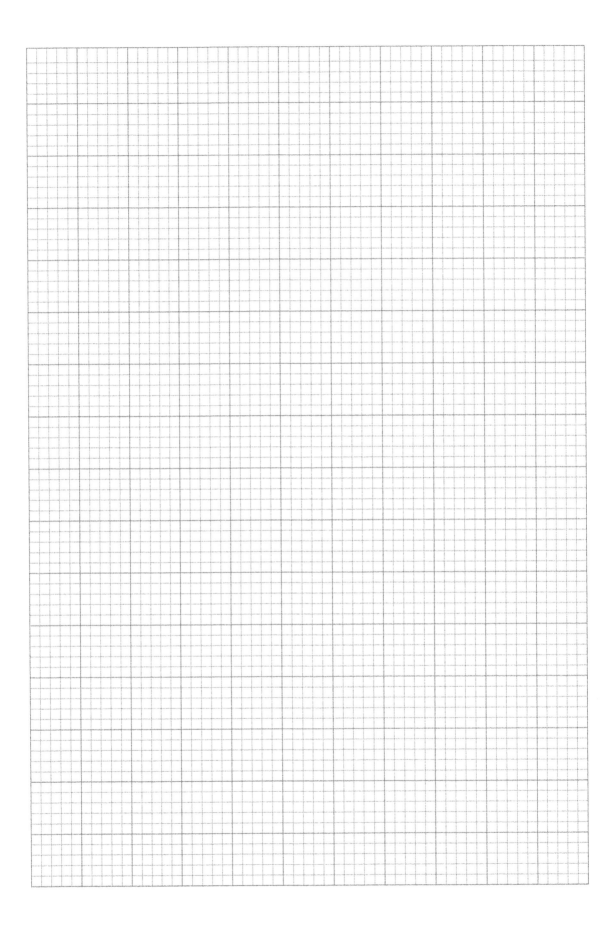

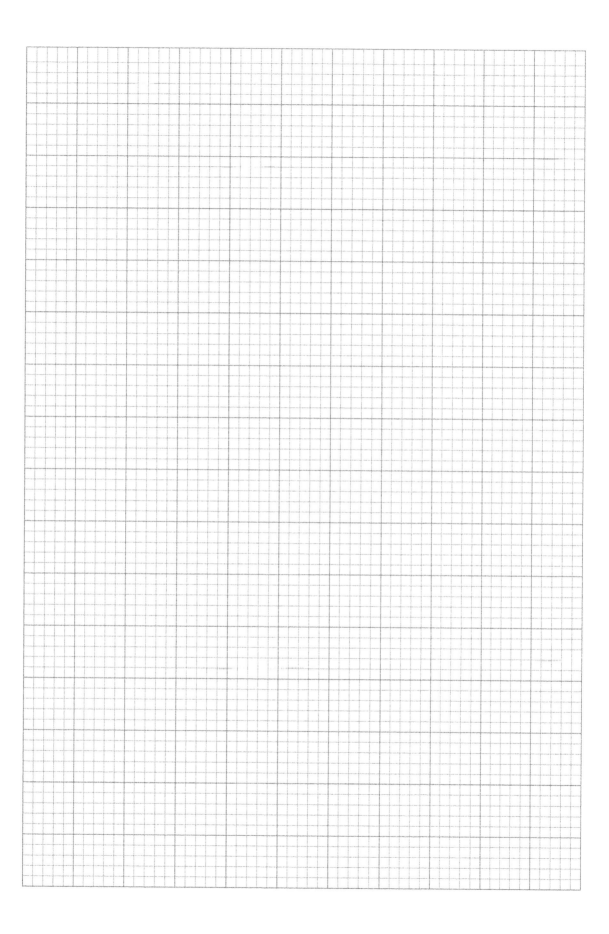

Notes

Notes

Notes